The Hidden Internet

The Nervous System of Civilization

Team Trex

Published by NobleTrex Press

For permissions and other inquiries, write to:
P.O. Box 3132, Framingham, MA 01701, USA

Contents

Introduction

One morning in 1858, Queen Victoria sent a message to President James Buchanan. It was a banal pleasantry, a mere ninety-eight words of diplomatic fluff, yet it sparked parades in the streets of New York City. Fireworks were launched, the City Hall was illuminated, and Tiffany's sold commemorative snippets of leftover cable. The excitement wasn't about what the Queen said; it was about the impossible physics of how she said it. For the first time in human history, a thought had crossed the Atlantic Ocean faster than a ship could sail.

That message took seventeen hours to transmit. A few weeks later, the insulation on the cable failed, the signal died, and the Atlantic went silent for another eight years.

Today, you can video chat with a stranger in Melbourne while sitting on a park bench in Madrid, and if the video stutters for half a second, you feel a surge of mild irritation. We have become metabolically dependent on a miracle. We treat the global delivery of information with the same casual expectation we apply to oxygen: we only notice it when it is gone.

We tend to speak of this digital existence in metaphors of weather and sky. We talk of "the Cloud," of data that floats, streams, and beams. We imagine the internet as an ethereal layer of reality, a ghost in the machine that exists everywhere and nowhere simultaneously.

This book is an argument that we are wrong.

The internet is not a cloud. It is a place. It is a heavy, thrumming, industrial beast built from iron, glass, copper, and concrete. It is the largest machine humanity has ever constructed, a sprawling nervous system that spans the globe, stitching together our economies, our governments, and our minds. And unlike the cloud, this machine is surprisingly fragile. It lives in specific buildings that you can walk past without noticing; it runs through cables that are vulnerable to shark bites, ship anchors, and backhoes; it relies on a politeness protocol between computers that was designed in an era when everyone on the network knew each other's first names.

To understand the modern world, we must look past the glowing screen in our hands and trace the wire back to the wall. We must strip away the sleek user interfaces and stare into the chaotic, messy, and magnificent engine room of civilization.

When you tap a link on your smartphone, you are triggering a chain reaction of industrial violence. To fetch a simple cat video, a signal does not float through the air (save for the first few meters to the cell tower). It dives into the ground. It travels as a pulse of light, trapped inside a strand of glass no thicker than a human hair, blasting through the darkness at two-thirds the speed of light.

This pulse must survive a journey through the bottom of the ocean, traversing the abyssal plain where the pressure is strong enough to crush a submarine. It must navigate the chaotic geography of the internet's routing tables, a digital bazaar where thousands of private networks barter and trade your data like a commodity. It must arrive at a fortress-like data center, perhaps in the humid heat of Singapore or the frozen plains of Luleå, Sweden, where it wakes a sleeping server. And then, it must do it all again in reverse, arriving back at your retina before your brain has time to register a delay.

The fact that this works at all is improbable. The fact that it works billions of times a second, for billions of people, is the single greatest engineering achievement of our species.

But this system is not just a glorified post office. As we will discover, the digital network has merged with the physical world to create a new kind of organism. We are no longer talking about a communication network; we are talking about a control system.

Consider the modern shipping container. As we explore the choreography of atoms, we will see that the movement of physical goods-coffee beans, semiconductors, insulin-has been completely packet-switched, modeled after the way data moves through the internet. The supply chain is now a physical manifestation of code. When the network stutters, ships stop. When the code breaks, the shelves go empty. We have built a civilization on the premise of "Just-in-Time" delivery, a philosophy that prioritizes efficiency over resilience, removing the buffers that once protected us from disaster. We have wired the logic of the internet into the gears of industry, creating a world where a software update in Texas can shut down a factory in Taiwan.

This convergence creates a paradox. We have never been more connected, yet our systems have never been more opaque. The complexity of this machine has surpassed the ability of any single human mind to fully comprehend it. It is an "accidental organism," evolving through a process closer to biology than architecture. As complexity scientists have long observed, complex adaptive systems-like ant colonies or the human brain-display emergent behaviors that cannot be predicted by looking at the individual parts. The internet is no different. It has moods. It has spasms. It creates feedback loops that can amplify a minor error into a global catastrophe, a phenomenon we will dissect when we look at the anatomy of a crash.

Why does this matter to you?

It matters because we are currently living through a transition as profound as the invention of the printing press or the steam engine. The nervous system of civilization is being tested. We are building this infrastructure on a planet that is becoming more volatile, subjecting our delicate glass threads and coastal data centers to the rising tides and fiercer storms of a changing climate.

Simultaneously, we are seeing the end of the internet's "age of innocence." The protocols that govern how networks talk to each other-such as the Border Gateway Protocol (BGP) and the Domain Name System (DNS)-were built on trust. They assumed that no one would lie about who they were or where they were located. Today, that trust is being weaponized. Nations and criminals are learning that they don't need to bomb a power plant to destroy it; they just need to hijack the routing tables that tell the plant's safety sensors where to send their data. We are witnessing the militarization of the map itself.

Yet, this is not a book about doom. It is a book about resilience.

One of the most surprising truths about the internet is that it is held together by people. We tend to imagine the network as automated and cold, but it is deeply, messily human. Behind every seamless video call and every instant transaction, there is a hidden army of engineers, technicians, and policy wonks who wake up at 3:00 AM when a pager goes off. They are the keepers of the flame, the ones who patch the servers, splice the broken cables, and negotiate the treaties that keep the traffic flowing.

In the chapters to come, we will descend into their world. We will follow the path of the physical infrastructure, starting from the crushing depths of the ocean floor where the "weight of the cloud" is most apparent. We will walk through the humming, sterile corridors of the "cathedrals of compute," the massive data centers that consume as much electricity as small cities, transforming energy into memory.

We will then learn the language of the machine. We will decode the handshake that allows a fridge in Ohio to communicate with a server in Zurich, and we will visit the "Bazaar at the Edge of the World," the chaotic economic marketplace where competing corporations agree to connect their networks for the greater good.

We will look at the system through the lens of biology, understanding how order emerges from chaos without a central captain. We will see how this digital nervous system drags the physical world along with it, orchestrating the movement of steel and wheat. And we will face the shadows-the crashes, the hacks, and the single points of failure that keep security researchers awake at night.

Finally, we will meet the people. We will explore the culture of "uptime" and the frantic, heroic efforts required to maintain the illusion of a frictionless world.

In 1934, T.S. Eliot asked, "Where is the wisdom we have lost in knowledge? Where is the knowledge we have lost in information?" Today, we might add: Where is the reality we have lost in virtualization?

To understand the future of our society, we must understand the ground it stands on. We must appreciate that the internet is not magic. It is a triumph of materials science, logic, and cooperative labor. It is a handshake across the world, sustained by voltage and vigilance.

The screen you may be reading this on right now feels like a portal to a different dimension. But if you look closely enough, you can see the iron, the glass, and the copper. You can feel the heat of the electrons. You can hear the hum of the civilization we have built.

Let's open the machine and see how it works.

Chapter 1

The Weight of the Cloud

The metaphor of the "cloud" suggests a floating, weightless abstraction, yet the internet is arguably the heaviest machine humanity has ever constructed. By tracing the physical journey of data through deep-ocean trenches, industrial cooling towers, and high-voltage substations, we reveal the massive, heat-generating infrastructure required to make the ephemeral feel permanent.

1.1 Threads Across the Abyss

On an ordinary afternoon, a photo you send, a video call you join, or a bank transfer you approve feels weightless. The word *cloud* encourages the illusion: information as mist, drifting wherever it pleases. The reality is narrower, sharper, and more fragile. Most of the world's international data does not travel by satellite. It travels through glass.

Not glass in the windowpane sense, thick and brittle, but a strand so thin that, held up to the light, it can disappear against the crease of your fingerprint. In that strand, pulses of laser light represent the ones and zeros of modern life. Those pulses cross oceans because the fiber is wrapped in an onion of protection: plastic insulation, steel strength members, copper for power, more steel for armor, and a final sheath designed to survive saltwater, pressure, abrasion, and time.

The result looks like an industrial cable you could imagine hoisting with a winch. At its core is something closer to a human hair.

This mismatch between the physical and the abstract is the point. The internet is not merely *connected*; it is *cabled*. Its grandest connections are not airy. They are laid on seabeds and buried in coastal sand, routed around mountain ranges of underwater topography, and threaded through narrow geographic choke points where ships anchor and fishing fleets work.

A useful way to picture a submarine cable is to imagine a long, continuous laboratory experiment that must operate unattended for decades. On land, if a component fails, a technician can drive to a cabinet and swap it out. Under the Atlantic, there is no cabinet. There is pressure, darkness, and an environment that corrodes everything that is not carefully chosen. So the cable is designed as a sealed, passive highway for light, with only a few crucial active devices interrupting the route: repeaters that boost the signal as it weakens with distance.

Light in fiber does not travel forever. Even in extremely pure glass, a tiny fraction of the signal is scattered and absorbed. Over hundreds and then thousands of kilometers, that loss accumulates. The solution is amplification: at intervals along the cable sit repeaters, metal cylinders engineered to survive the deep sea, whose job is to strengthen the optical signal without converting it back into electronics. Inside many modern repeaters is an erbium-doped fiber amplifier, often shortened to EDFA. The name is less important than the idea: the cable carries not only the data signal but also electrical power, and that power runs lasers that "pump" a special fiber segment so it can amplify the passing information-carrying light. The internet's longest conversations are therefore sustained by a chain of underwater devices, each one quietly fighting the gradual dimming of photons.

None of this is romantic when you see it up close. Submarine cables are industrial infrastructure, designed and financed with the same unglamorous realism as pipelines and railways. Their most delicate element is hidden, because it must be. The ocean is not a neutral medium. Near shore it is busy, crowded with anchors, trawls, dredges, and storms that rearrange the seabed. Farther out it is stable in a different way: less human disturbance, but immense pressure, and a remoteness that turns any repair into an expedition with paperwork, specialized vessels, and long waits.

A historical jolt helps here, because humanity has been surprised by the ocean before. In 1858, the first transatlantic telegraph cable briefly connected Ireland to Newfoundland. For a few weeks, messages that once took days by ship crossed the ocean in minutes. Celebrations erupted; newspapers announced a new era; dignitaries exchanged greetings. Then the cable failed. Engineering mistakes, over-ambitious voltages, and the difficulty of understanding a system that could not be inspected along most of its length combined into silence. The dream did not die, but it did acquire a price: the price of learning, in public, that the seafloor is not a friendly place for fragile signals.

The modern fiber-optic era is the descendant of that lesson. If the nineteenth century discovered that oceans could carry information, the late twentieth century discovered how to carry *enough* information to matter economically. A single fiber can carry many channels of light at different wavelengths, like a multi-lane highway built inside a transparent thread. The breakthrough is not just the purity of the glass, but the trick of *multiplexing*—packing many separate data streams into one physical path by using different colors of infrared light, then unpacking them at the far end. When you hear that a cable system can carry tens of terabits per second, it is tempting to imagine brute force: thicker cables, bigger equipment. Much of the increase is instead intellectual compression: clever ways to encode more information per unit time, plus an arms race of reliable undersea amplification and terminal equipment.

And yet, the cable itself remains, stubbornly, a physical object. That means its failures are physical too.

One of the most instructive cable failures in recent memory did not occur in the middle of the ocean but at the edge of it, in exactly the zone where a modern map of the internet begins to look like a map of human activity. In January 2022, the volcanic eruption near Tonga severed the nation's primary submarine cable, cutting off most communications with the outside world. The cause was not a ship's anchor or a fishing net but a geological event powerful enough to reshape the seafloor. Satellite links provided limited backup, but the bandwidth and latency were no substitute for a fiber connection. The episode was a reminder with a human face: connectivity is not a background condition. It is a lifeline, and lifelines can snap.

Many disruptions are more mundane. A ship drops anchor where it should not, the anchor drags, and the dragging becomes a plow. A trawler's gear bites into the seabed with the indifference of a machine built to withstand force, and the cable, unless deeply buried, is simply another obstacle. The idea of sharks biting cables is famous enough to be a joke, and there have been incidents, but the more typical offenders are our own tools. This is why the cable's coastal segment often looks different from its deep-sea segment. In shallow water, where risk is highest, the cable is thicker, more heavily armored, and frequently buried. In deep water, where the seabed is quieter and laying long distances is expensive, the cable can afford to be lighter.

This engineering logic produces a geography of vulnerability: the parts of the system closest to people are often the parts most likely to fail. The paradox is that the most consequential routes are often forced close to people. Landings must occur somewhere. They tend to cluster where there is existing infrastructure, stable politics, and access to power and backhaul fiber on land. As a result, coastlines become the internet's ports and harbors, and like any port, they are a mix of commerce, congestion, and risk.

A submarine cable system is not just the cable. It is also the landing station: a secured building where the undersea fiber terminates, is amplified and converted, and then hands off traffic into terrestrial networks. Landing stations are less famous than data centers, but they have a similar character: controlled access, careful power, redundant equipment, and an uneasy relationship with the public imagination. People accept, abstractly, that information must go somewhere, but the idea that their country's digital umbilical cord comes ashore at a specific beach can feel too concrete, too targetable. In some places, this has become a national security concern, because cables are both vital and difficult to defend comprehensively. You cannot post a guard along thousands of kilometers of seabed.

Even without malice, the ocean has a way of distributing stress. Submarine landslides can occur when sediment on a slope gives way, triggered by earthquakes or storms. The sliding material can move fast enough to sweep cables along like threads caught in a current. When that happens, failure is not a cut so much as a violent relocation: the cable is displaced, tensioned, snapped, sometimes in multiple places. The seabed itself becomes a moving machine.

For everyday users, the most interesting part of these failures is not the break but the repair. Repairing a cable is a choreography that exposes just how physical the internet remains.

First, the fault must be located. This is done by measuring how light reflects back along the fiber, using techniques that treat the cable like a long optical instrument. Then a specialized cable ship is dispatched.

These vessels are not generic workboats. They carry massive drums of cable, heavy lifting equipment, and crews who know how to handle kilometers of line without twisting or damaging it. They must also navigate logistics: permissions to operate in territorial waters, coordination with local authorities, weather windows, and the simple fact that the global fleet of such ships is limited. When a break occurs near a crowded coastline, the ship may be competing for space with fishing fleets and commercial traffic. When a break occurs far offshore, the problem is distance and time.

Once on site, the ship lowers a grapnel to snag the cable from the seafloor, lifts it aboard, cuts out the damaged segment, splices in new cable, tests the optical performance, and lowers it back down. If the cable is meant to be buried, a plow or remotely operated vehicle may be used to trench it into the seabed. The operation looks, from above, like mending a thread with gloves on while standing on a moving platform.

Splicing fiber is a delicate art performed inside an environment that must be controlled even on a ship. The goal is to align two glass cores so precisely that light passes through with minimal loss. Modern splices are astonishingly good, but the work remains the opposite of rugged. It is the careful rejoining of hair-thin glass so that global traffic can resume.

This combination—a repair process that is at once heavy and delicate—is the signature of submarine infrastructure. It is also why redundancy matters. The global network is not a single cable but a mesh: multiple routes between continents, multiple landings, multiple operators. When one route fails, traffic can often be rerouted. The rerouting happens at higher layers of the network, through agreements and protocols that decide which path is available and economical. To the user it looks like resilience; to the engineer it looks like a system that anticipates failure by assuming it is normal.

But redundancy is not evenly distributed. Some regions have many cables and many alternate routes; others have only a few. Islands and smaller economies can be especially exposed: one or two main cables, perhaps a backup with less capacity, and then satellites for emergencies. Even in well-connected regions, geography concentrates risk in places where many cables share corridors. The seafloor has its own highways: relatively shallow, relatively stable routes that avoid deep trenches and seismic zones, and which are also often near busy coastal waters. The infrastructure's efficiency and its fragility can therefore be intertwined.

A cross-section of the cable makes the point more vividly than any metaphor, because it shows how much engineering is devoted to protecting something almost comically small.

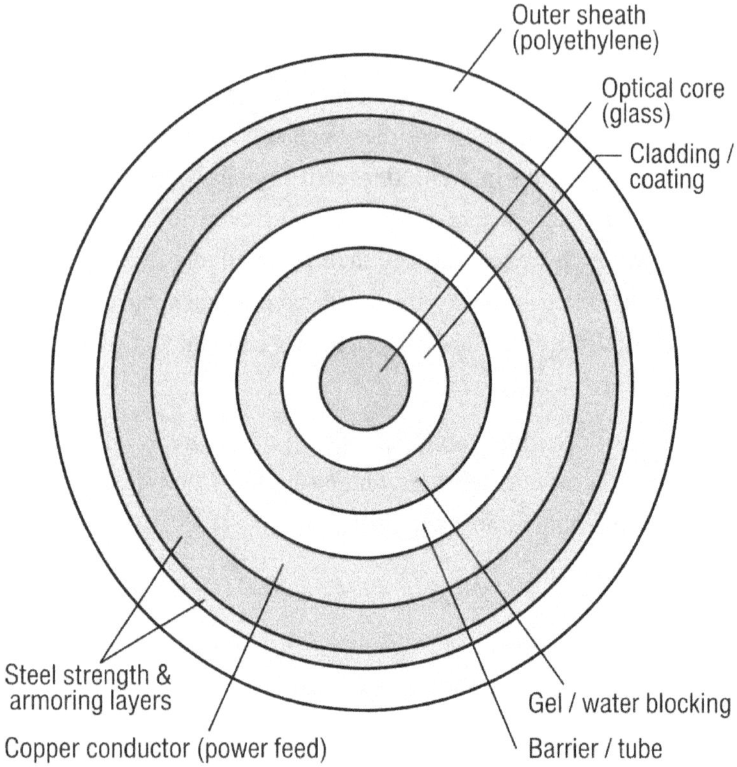

Look again at the proportions: a bold dot of red at the center for the core, and then layer upon layer devoted to keeping that dot intact. The cable is a kind of reverse skyscraper: the valuable space is not the exterior but the tiny interior through which light must pass cleanly for years.

There is another, subtler fragility here: the cable does not merely transmit light; it is part of an electrical system too. The repeaters need power, and power must be supplied from the ends. That means the entire system, spanning an ocean, behaves like a long electrical circuit under the sea. The engineering must account for voltage, current, insulation, and failure modes that would be mere inconveniences on land but are intolerable when the equipment is inaccessible. The undersea repeaters are designed for extraordinary reliability because a single failed repeater can degrade or disable vast capacity. A cable system is therefore built around a question that sounds almost philosophical: what kind of machine can you deploy where you cannot repair it easily?

The answer, historically, has been to make the machine as simple as possible in the hostile environment and concentrate complexity at the ends. Much of the cleverness lives in the terminal stations, where signals are generated, multiplexed, routed, and monitored. Under the ocean, the system is comparatively austere: fiber, repeaters, and protective layers. This division of labor echoes a broader pattern in infrastructure: put the delicate intelligence where humans can reach it, and build the rest to endure.

That endurance is not only technical. Submarine cables cross jurisdictions, touch rival economies, and connect allies and competitors with the same indifferent efficiency. They require permissions, treaties, and a practical culture of cooperation among operators who may compete in the market but share the same ocean. Cables are private investments that behave like public utilities, and that tension shows up whenever something goes wrong. When a fault occurs near a coast, the repair ship's arrival can feel, to local authorities, like the arrival of a critical-care team for the nation's circulation.

The most unsettling fact about submarine cables is not that they break. Everything breaks. It is that they break in ways that reveal how much of modern life depends on a small number of physical paths. The ocean is wide, but the routes between the places we care about are narrower than we like to admit. If the cloud has a weight, part of it is the weight of this dependence: commerce, family, diplomacy, entertainment, and emergency services all sharing, for long stretches, the same thin strands of glass.

The seabed does not celebrate our dependence or resent it. It simply sits there, dark and cold, while light flickers through protected hair-thin cores. Above, ships move, storms pass, fishing gear drags, tectonic plates shift. The internet persists not because it floats free of the world, but because it is lashed to the planet with threads engineered to survive everything except, occasionally, the ordinary force of contact.

1.2 Cathedrals of Compute

The cable on the ocean floor carries light. At the far end of that light is a building that looks, at first glance, like any other industrial box on the edge of a highway: low, wide, windowless, fenced. The reason it has no windows is not secrecy, though it is guarded. It has no windows because it is not built for people. It is built for heat.

A data center is an agreement with physics. Every useful computation turns electrical energy into order *inside* a machine and disorder *outside* it. The language of physics calls that disorder entropy, and its habit of spreading is summarized by the Second Law of Thermodynamics. A data center does not repeal that law; it pays it, continuously, in the currency of cooling systems, fans, pumps, heat exchangers, and backup power. When someone says "the cloud," they are naming an industrial heat engine whose main byproduct is warm air and, in some designs, warm water.

It is easy to underestimate the problem because the machines are small. A server is a metal drawer you can slide out with one hand. A rack is a tall cabinet. A row of racks looks like a corridor in a library, except the books hum. Nothing about the furniture seems to demand an entire building's worth of engineering. The trick is density. A modern rack can draw tens of kilowatts, and a row of racks can concentrate energy consumption into a footprint smaller than a suburban living room. A toaster feels hot because it is pushing a kilowatt into a small space. A row of AI-focused racks can be dozens of toasters stacked shoulder-to-shoulder, except the toasters must operate reliably, without smoke, for years, while doing precise work.

The oldest story in computing is the story of shrinking: computers that once filled rooms now fit in pockets. Data centers are the counter-story. Computing has shrunk in *devices* and grown in *places*. The shrinking created demand; the demand created concentration; the concentration created buildings that behave like power plants in reverse. A power plant takes heat and turns some of it into electricity. A data center takes electricity and turns almost all of it into heat.

That near-total conversion surprises people. It sounds like an insult to the industry, as if computing were wasteful by nature. In fact it is the opposite: digital electronics are astonishingly good at their job, but their job is performed by controlled currents and voltages in tiny transistors, and every controlled current leaks some energy as heat. Even the light in optical transceivers ends up as heat once it has done its work. The law is not "some energy becomes heat." The law is: in the end, *all* of it does. The only choice is where the heat appears and how quickly you can remove it.

This is why a data center's architecture begins with airflow, not processors. Walk through one and you are likely to feel something oddly theatrical: the cold is too cold, the noise too loud, the sense of movement too constant. You are in a building designed to make air behave.

The simplest way to keep servers cool is to blow cool air across their components and carry the warmed air away. But if you do that casually, the warm exhaust mixes with the cool intake, and the servers begin to "breathe" their own waste heat. Temperatures creep up, fans spin faster, energy use rises, and reliability falls. At scale, that mixing becomes the enemy. The modern response is a piece of choreography called *hot aisle / cold aisle containment*. Racks are arranged so that the fronts of servers face one corridor (the cold aisle) and the backs face another (the hot aisle). Cool air is supplied to the cold aisle; servers pull it in; hot air is expelled into the hot aisle; and then that hot air is returned to cooling units that remove its heat before sending the air back cold again.

The arrangement sounds trivial until you see how aggressively it is enforced. In many facilities, the cold aisle is physically contained with doors and a roof, like a tunnel. The goal is not comfort; it is separation. Keep the cold air cold. Keep the hot air hot. Mixing is waste.

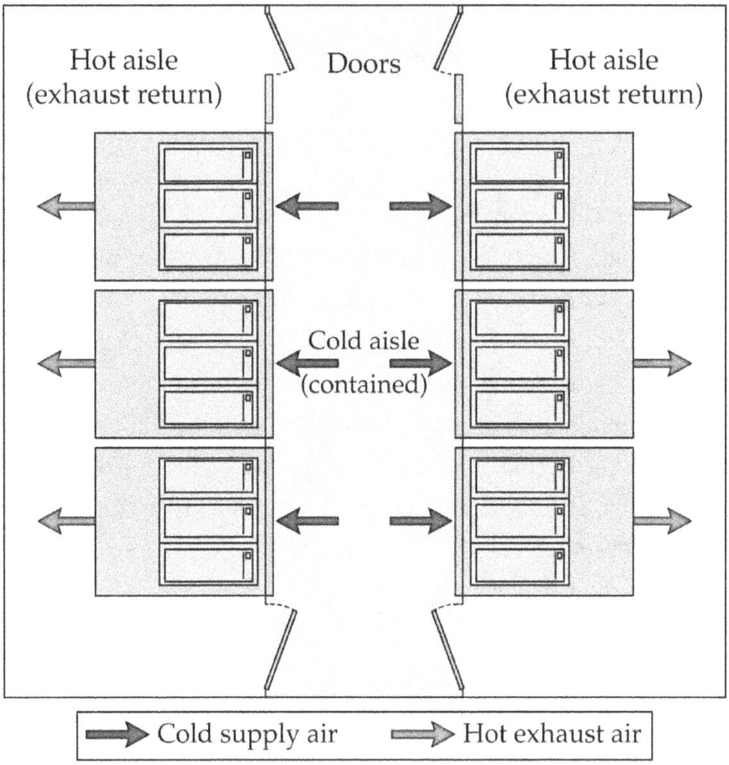

Once you notice the logic, you start seeing it everywhere. The facility is a sequence of separations: electrical rooms separated from wet cooling equipment; battery strings separated from human traffic; network rooms separated from loading docks; cold air separated from hot air; inside separated from outside. The popular image of a data center is a room full of blinking lights. The operational reality is a building designed to prevent small failures from becoming shared failures.

Some of that prevention is almost comically physical. A server does not simply compute; it gulps air. Fans pull air through narrow channels over CPUs, memory modules, and power converters, then push it out the back. If the intake air is even a few degrees warmer than planned, those fans spin faster. Faster fans mean more electricity devoted to moving air rather than doing computing, and the fan motors themselves add more heat. At scale, tiny inefficiencies multiply into large bills and uncomfortable margins. This is one reason operators talk obsessively about *PUE*, or Power Usage Effectiveness. PUE is a ratio: total facility power divided by IT power (the power going to servers, storage, and networking gear). A perfect, physically impossible data center would have a PUE of 1.0, meaning every watt goes to computing and none to overhead. Real facilities do better or worse depending on climate, design, and how hard the servers are being driven. Uptime Institute's surveys in the mid-2020s suggested that average PUE had stopped improving quickly, a hint that easy gains had been captured and further improvements required deeper redesign rather than simple best practices.

PUE is a blunt instrument, but it captures an important truth: the cooling system is not a side feature. It is part of the computer.

That becomes clearer when the workload changes. For decades, much of the industry assumed a certain kind of server: general-purpose machines with moderate power draw, cooled comfortably by air, with some variability but not extreme spikes. The rise of accelerated computing—specialized chips designed for machine learning and other parallel tasks—has complicated the picture. These systems can be astonishingly powerful, and power-hungry in a way that stresses the old assumptions. When a rack's heat output rises, the airflow required to cool it rises too, and airflow is not free. Push enough air through a narrow space and you meet limits: noise, pressure drop, fan reliability, and the simple fact that hot air wants to rise and spread.

This is where the phrase "battle the Second Law" stops being a clever line and becomes a design constraint. Entropy does not just increase in the abstract; it shows up as a hot spot behind a rack, a failed fan, a tripped breaker, a CPU throttling itself to avoid overheating. The operator's job is to keep a carefully bounded pocket of low entropy inside the servers: stable temperatures, clean power, predictable humidity, reliable timing.

The historical anecdotes of computing are full of machines that failed because they could not get rid of heat fast enough. Early mainframes were cooled with elaborate air conditioning, and their rooms became status symbols: modern temples with raised floors and controlled climate. Those raised floors were not only for hiding cables. They were a way to distribute conditioned air like water in plumbing, delivering it to vents in front of cabinets. The raised floor became an icon of "the computer age," but it was also a confession: computation creates heat, and heat must be carried away.

Hyperscale facilities—the vast data centers built by the largest cloud providers—have taken that confession and industrialized it. They have standardized designs, optimized supply chains, and tuned every part of the airflow path. The aisles are arranged with geometric discipline. Sensors monitor temperature and pressure at many points. Control systems adjust fan speeds and cooling output in response to demand. In a well-run site, the cooling plant and the IT load are in constant conversation: not a human conversation, but a tight feedback loop of measurements and automated responses.

The war against heat is also a war against time. Electronics age faster at higher temperatures. Solder joints fatigue. Capacitors degrade. Even if nothing fails catastrophically, running hotter can shorten the useful life of equipment, and at scale that becomes a replacement schedule, a supply chain dependency, and a financial plan. In other words, thermodynamics is not only about keeping the servers alive today. It is about keeping them economically alive three years from now.

Cooling, however, has its own costs and side effects. One of the most contentious is water. Many traditional cooling systems rely on evaporative cooling, where water is used to carry away heat as it changes phase from liquid to vapor. Evaporation is extraordinarily effective, and in dry climates it can be efficient in energy terms. But it consumes water, and it consumes it in places where water may already be scarce. Communities sometimes learn about nearby data centers not from the internet's speed but from debates at local planning meetings about water usage and heat discharge.

The industry has begun to respond with alternative designs, including closed-loop liquid cooling that aims to reduce or eliminate evaporative water loss. Microsoft's August 2024 announcement of a "zero water" cooling design for certain data centers is part of this broader shift in ambition: treat water not as an invisible utility but as a scarce resource that must be budgeted and, where possible, avoided. Whether any specific claim holds for every site and climate is less important than the direction of travel. The cloud is being asked to account not only for electricity but for its shadow resources.

Electricity, meanwhile, has become too large to ignore. When the U.S. Department of Energy discussed data centers in 2024, it pointed to figures that startled many readers: data centers consuming about 4.4% of total U.S. electricity in 2023, with scenarios reaching roughly 6.7% to 12% by 2028, depending on growth and efficiency. Globally, the International Energy Agency's 2025 analysis described a world in which data center electricity consumption could roughly double by 2030. These are not marginal loads. They are the kind of loads that change how grids are planned and how communities argue about land use.

Yet the inside of the building is not only cooling plant and servers. It is also the engineering of continuity: what happens when power fails, when a cooling pump breaks, when a switch dies. Data centers are "fortified" not because they expect war, but because the combination of scale and dependence makes ordinary failures unacceptable. Backup generators, batteries, redundant power feeds, multiple network paths, fire suppression systems designed not to destroy electronics, and careful maintenance regimes are the practical response to a simple social fact: when a large provider goes down, thousands of other businesses go down with it.

A popular misunderstanding is that a data center is a giant hard drive, a warehouse where "the internet" is stored. Storage matters, but the livelier truth is that a data center is a place where computation is performed continuously on behalf of others. A search query, a map request, a streaming recommendation, an AI-generated paragraph, a payment authorization: these are not retrieved like books from shelves. They are computed by software running on many machines, orchestrated so that the user sees a coherent service rather than a swarm of processes.

That orchestration has an unexpected thermodynamic implication. In a library, you can leave most shelves untouched and energy use stays flat. In a data center, even "idle" machines often consume meaningful power, and machines are kept ready because users demand low latency: the time delay between request and response. Low latency is not merely a preference; it changes what can be built. If a facility must respond in milliseconds, it must be close enough to major populations, and it must keep resources warm and waiting. The cloud's promise of immediacy is paid for in baseline power draw and the continuous circulation of cooling air.

Engineers are attempting to reclaim some of that cost by increasing utilization: running servers closer to their capacity so that the overhead is amortized over more useful work. But higher utilization often means higher sustained heat output, which pushes cooling systems harder. The dream is not to avoid thermodynamics; it is to spend the thermodynamic budget more wisely.

There is something almost architectural about the way these buildings express their purpose. Cathedrals were built around a kind of immaterial aspiration, but their design was shaped by gravity: stone arches, buttresses, load paths. Data centers are built around immaterial work too, but their design is shaped by heat: airflow paths, containment, redundancy, and the relentless need to push disorder out of the building as fast as the servers create it.

Stand in the cold aisle long enough and you can feel the invisible geometry. The air presses gently against your skin, moving with intention. Behind you, separated by a few meters of metal and a disciplined arrangement of doors and panels, is the hot aisle, where exhaust air collects like breath. Nothing about the scene suggests the content of the computation—love letters, payroll, emergency calls, machine learning models—and everything about it suggests the price. The cloud is not lighter than air. It is heavier than heat, and it must be carried away, hour after hour, into the open world.

1.3 The Heartbeat of Voltage

A packet of data does not weigh anything you can feel, but it has a cost that shows up on a power bill. The ocean-spanning fibers are dark without electricity at their landing stations and repeaters. The racks of servers are inert metal without power converters feeding their chips. Even the most abstract act online—refreshing a page, asking a model for a sentence—begins as a physical transaction between a generator and a load, mediated by wires, transformers, and control rooms.

The internet, in practice, is a specialized subset of the electrical grid. Not because the grid contains it like a folder contains a document, but because the internet's organs are designed around a single assumption: voltage will be there, continuously, within narrow tolerances. Every other reliability story in computing sits downstream of that assumption.

The grid is easy to romanticize as a vast battery, but it is not a battery. It is closer to a moving tightrope act: supply and demand must be balanced almost instantaneously. Electricity is difficult to store at scale, so most of what is consumed at a given moment must be generated at that moment. That balance is not a gentle equilibrium. It is an engineered compromise among machines with inertia, markets with prices, weather with moods, and a society that expects the lights to stay on.

Frequency is the grid's heartbeat. In North America it is nominally 60 Hz: the alternating current oscillates sixty times a second. When supply falls behind demand, frequency sags; when supply exceeds demand, frequency rises. Those changes are usually small, but they are meaningful. Many devices can tolerate some drift, but large deviations can damage equipment and, more importantly, signal that the system is sliding toward instability. Grid operators watch frequency the way an intensive care unit watches a patient's pulse.

The internet is unusually sensitive to that pulse, not because routers and servers require perfect sine waves, but because they require *predictable* power: stable enough that power supplies can regulate it, and stable enough that the cooling systems can keep up. The dependence is layered. If grid power wobbles, data centers switch to batteries; if an outage persists, they start generators; if fuel deliveries are disrupted or generators fail, the building heats up, and computation stops. The illusion of an always-available cloud is built on a chain of contingencies, and the first link is the public electrical system.

The grid, for its part, has come to depend on the internet in a different way. Modern power systems are monitored and controlled by networks of sensors and industrial computers. The umbrella term for these systems is SCADA: Supervisory Control and Data Acquisition. In plain language, SCADA is the nervous system inside the grid's machinery. It measures voltages and currents, opens and closes breakers, sets transformer taps, and coordinates the switching that keeps power flowing around faults. It is not the consumer internet in the way a phone is, but it is networked, and it increasingly relies on digital communications that share technologies, vendors, and vulnerabilities with the wider world.

That creates a recursive fragility: the internet depends on the grid for power, while parts of the grid depend on digital networks for coordination. When everything works, this mutual dependence looks like efficiency. When something breaks, it can become a loop.

A concrete example came into public consciousness in December 2015, when a cyberattack on Ukraine's power system led to outages affecting hundreds of thousands of customers. The event became notorious not only because the lights went out, but because it offered a glimpse of how digital control systems can be manipulated. It was not the first major outage in history, and it was not purely "an internet problem," but it reminded the world that electricity has become software-mediated. In the years since, grid operators and cybersecurity teams have treated the boundary between operational technology and information technology as a frontline rather than an accounting category.

The everyday physics of the grid can feel distant until it collides with ordinary online life. Consider what a data center looks like to a utility planner. A neighborhood grows gradually: a few houses, then a development, then a shopping strip. A hyperscale data center can arrive as a single project with the appetite of a small town. The planners must decide where the power comes from, what transmission and distribution upgrades are required, who pays for them, and how confident they are that the demand will actually materialize.

This last question has become surprisingly thorny. In the mid-2020s, utilities and grid operators began to confront what journalists sometimes called "phantom" data centers: load requests that appeared in interconnection queues or planning documents but were speculative, duplicative, or contingent on financing and permits that might never arrive. From the utility's point of view, building infrastructure for imaginary demand is not merely wasteful; it can shift costs onto other customers. From the data center developer's point of view, securing a place in the queue can be rational when lead times are long and competition is fierce. The tension is not primarily moral. It is a clash between two time horizons: the grid's slow construction schedule and the technology sector's rapid iteration.

Northern Virginia, particularly the cluster often called "Data Center Alley," has become a symbol of this collision. A region that once looked like ordinary suburbia now hosts dense concentrations of computing facilities because of proximity to network backbones, major population centers, and an ecosystem of vendors. The local grid has had to treat these facilities not as background commercial load but as a defining feature of the future. When a single region attracts many large data centers, the grid operator must consider correlated risk: heat waves that drive both air conditioning demand and data center cooling demand; transmission constraints that become binding at the same time; and the possibility that a disturbance could cascade through tightly loaded infrastructure.

Grid operators talk about adequacy and reliability in a language that sounds calm precisely because the underlying reality is not. Adequacy means there is enough capacity, in principle, to meet expected peak demand. Reliability is broader: the ability to withstand disturbances, faults, and the loss of individual components without widespread outage. Reports from regulators in the mid-2020s often struck a cautious tone: the system could be adequate under normal conditions while still facing tightening margins. Those margins matter because they define how much surprise the grid can absorb. A sudden generator outage, a transmission line fault, an unexpected surge in demand: these are not rare events. They are the grid's daily weather. Tight margins mean fewer options and faster decisions.

The internet's dependence on those margins is not theoretical. When the grid is strained, operators may ask large customers to curtail load. Some industrial facilities can do this easily: a factory can pause a process, or a smelter can ramp down. A data center can sometimes reduce non-urgent computing, shift workloads to other regions, or rely more heavily on on-site generation, but these options have limits. Many online services are designed to be geographically redundant, yet that redundancy depends on network capacity, on power availability elsewhere, and on contractual arrangements that allow shifting. When curtailment becomes a serious possibility, it forces an uncomfortable conversation: which parts of the cloud are essential, and which are merely convenient?

Inside data centers, the answer has historically been to behave as if the grid is unreliable by default. That is the logic behind redundant power feeds, batteries, and generators. In the language of facility design, redundancy is not one feature but an entire philosophy. Uptime Institute's Tier classifications, for example, describe different levels of fault tolerance and maintainability, including whether there are multiple independent distribution paths and redundant components. The tiers are not a guarantee of perfection, but they reflect a worldview: a serious facility is designed so that routine failures do not interrupt service.

Yet redundancy is not free, and it is not purely private. Diesel generators require fuel storage and emissions permits. Battery systems require supply chains and fire safety engineering. Multiple utility feeds require grid infrastructure that must be built and maintained. In this sense, the private reliability of the cloud is entangled with public planning decisions. A data center can pay for some of its own fortress walls, but it cannot privately build the regional transmission network that feeds it.

Thermodynamics returns here in an unexpected form. The servers' heat must be removed; the cooling systems require electricity; and the grid must generate that electricity without destabilizing frequency and voltage. The cloud's physical weight is therefore not only a matter of megawatts; it is a matter of *when* those megawatts are demanded. Peak demand is the grid's critical moment. If many data centers in a region draw heavily during the same hours as everyone's air conditioners, the system's stress is concentrated. Operators respond with pricing signals, demand response programs, and, increasingly, negotiations with large customers about flexibility.

Flexibility is a subtle promise. A data center operator may say, truthfully, that some workloads can move in time or space. Training a large machine learning model might be postponed a few hours, or shifted to another facility hundreds of miles away. But not everything is flexible. User-facing services must respond instantly. Business contracts may require certain uptime. Latency considerations make some workloads geographically sticky. The grid would like load to behave like a dimmer switch; the internet often behaves like a heartbeat monitor that must not flicker.

The dependence between grid and internet becomes most visible during extreme events. A hurricane can knock out transmission lines and also flood low-lying facilities. A wildfire can trip lines and also degrade air quality, complicating cooling. A winter storm can freeze fuel supplies for generators and also increase heating demand. In February 2021, the Texas power crisis showed how multiple layers of infrastructure can fail together under stress: generation outages, fuel constraints, and cascading consequences for households and businesses. Data centers in such situations may ride through short disturbances on batteries, then on generators, but the broader lesson remains: resilience is as strong as the most constrained shared resource.

There is a tendency to treat the grid as a background service like water from a tap. That metaphor breaks down because electricity is more like a conversation than a delivery. The consumer speaks by switching on load; the grid responds by adjusting generation and routing power through available lines; control systems speak back through prices, demand response calls, and protective relays that trip when limits are exceeded. The conversation is continuous, and the internet is one of its loudest new speakers.

This is why debates about the cloud's growth are often, at heart, debates about governance. Who gets to claim scarce capacity? Who pays for upgrades? How are forecasts made, and what happens when they are wrong? In regions with organized wholesale markets, these questions are filtered through institutions with acronyms: RTOs and ISOs, commissions, reliability councils. Their decisions can sound bureaucratic, but they determine whether the internet's industrial backbone expands smoothly or collides with hard constraints.

There is also an ethical dimension that hides inside technical language. A megawatt diverted to one purpose is a megawatt not available for another. When electricity is abundant, the tradeoff is invisible. When the system is tight, it becomes political. Communities may welcome data centers for jobs and tax revenue, or resent them for land use, noise, diesel emissions, water concerns, and the feeling that local resources are being converted into remote digital convenience. None of these reactions are irrational. They are the human face of the Second Law: the benefits of computation are distributed widely, while the heat, the wires, the substations, and the fuel tanks are located somewhere specific.

The phrase "the cloud" does not easily accommodate a substation, yet the substation is the cloud's front door. Step-down transformers convert high-voltage transmission into the medium voltages that serve industrial customers, and then into the low voltages that feed server power supplies. Those power supplies rectify alternating current into direct current and regulate it into rails precise enough for microelectronics. Each conversion loses some energy as heat. The losses are small in percentage terms, but the quantities are large, and they add to the building's thermal burden. The internet's nervous system runs on electrons that have been stepped, transformed, switched, and conditioned many times before they become a clean, steady supply inside a server.

In the end, the internet is not a parallel world made of pure information. It is an electrical phenomenon given meaning by protocols and people. Its most glamorous products are weightless; its enabling condition is profoundly material. Somewhere, at the moment a message is delivered, a generator's torque is resisting a load, a control room is keeping frequency near 60 Hz, a transformer is humming, and a cooling system is carrying heat away so that the next message can be delivered without hesitation.

Chapter 2

A Handshake Across the World

Imagine a crowded room where every guest speaks a different language, yet a flawless conversation ensues. This chapter dismantles the invisible diplomatic protocols that allow billions of strangers-silicon and human alike-to exchange data without misunderstanding, revealing the rigid etiquette essential to the chaos of the open web.

2.1 The Polite Conversation

Anyone who remembers the landline era also remembers the small humiliation of the *busy signal*. You dialed, you waited, and a tone informed you that the path between you and the other person was unavailable. Nothing was *wrong* with your telephone. Nothing was necessarily wrong with theirs. The network was simply full in the particular way it could be full: a limited set of dedicated lines was already spoken for. The network was not a marketplace of messages; it was a set of corridors. If all the corridors were occupied, newcomers stood outside.

That sound captured the logic of circuit switching. A telephone call did not merely *use* the network; it *became* a temporary piece of the network. For the duration of the call, the system carved out an end-to-end circuit, a reserved route through a chain of switches, and kept it exclusively for you. The advantage was psychological as much as technical: once the circuit existed, your voice had a private lane. It arrived in order, in real time, with a steadiness that felt like physical presence.

The cost of that steadiness was waste and fragility. Waste, because speech is intermittent: in ordinary conversation there are pauses, hesitations, laughter, thinking. A reserved line cannot be leased to someone else for those pauses; the emptiness is still yours. Fragility, because the whole arrangement depends on the continuity of the route. Break any link in the chain and the call collapses. A circuit-switched network treats a conversation as a single object that must be carried intact from one end to the other. It is a beautiful idea, and it works well when the main thing the network carries is the *illusion of immediacy* between two human beings.

Now consider what the modern internet carries: not only voices, but photographs, web pages, software updates, medical records, financial transactions, and the tiny background chatter of devices asking for time, location, permission, and identity. Some of this information is urgent; much of it can tolerate brief delay. Some arrives in bursts; some flows steadily. Treating all of it as if it were a continuous telephone call would be like insisting every letter be delivered by a dedicated courier who walks from your house to the recipient's, never stopping, never sharing the road.

Packet switching replaced the corridor with the postal system.

In a packet-switched network, information is chopped into small pieces called *packets*. Each packet is a compact envelope: it contains a fragment of the payload (the actual data) plus addressing information that helps the network decide where to send it next. Packets from the same message do not have to travel together. They do not have to take the same route. They do not even have to arrive in order. The network's job is no longer to preserve a continuous end-to-end circuit. Its job is to move packets hop-by-hop, making local decisions, sharing links among many users, and relying on the endpoints to reconstruct meaning from the arriving fragments.

The trick is that packet switching takes for granted something circuit switching tried to avoid: unpredictability. Links get congested. Packets collide, queue, and wait. Some are dropped. Routes change. Hardware fails. In exchange for accepting this messy reality, packet switching can do something circuit switching struggles with: it can keep functioning when parts of the network are broken, overloaded, or under attack, because traffic can flow around damage rather than depending on a single reserved path.

A familiar historical anecdote helps make the mindset shift concrete. During the Cold War, engineers and planners worried about communications surviving disasters, including attacks on infrastructure. A circuit-switched system has an obvious weakness: destroy a key switching office or trunk line and entire regions may be cut off. Packet switching does not magically make cables indestructible, but it changes the failure mode. If the network is a mesh of possible routes, and if messages are not bound to a single end-to-end circuit, then losing a node or link can be handled as a routing problem rather than an existential catastrophe. Messages can be re-routed, perhaps more slowly, perhaps with reduced capacity, but without requiring the entire system to agree on and preserve one fragile chain.

The internet's resilience is often described as if it were a heroic quality, but it is closer to a mundane virtue: the ability to keep going when conditions are imperfect. The designers of packet-switched protocols assumed imperfection from the beginning. Instead of engineering a world where nothing is lost, they engineered a world where loss is normal and recoverable.

The conversation metaphor matters because TCP/IP does not merely move data; it establishes *etiquette* between strangers. The core internet protocol, IP, is deliberately unsentimental. It provides addressing and best-effort delivery: each packet is sent, and the network does what it can, but no promise is made that the packet will arrive, arrive once, or arrive in order. IP is like tossing bottles into a river that usually flows the right way.

The politeness enters with TCP, the Transmission Control Protocol. TCP sits above IP and turns this unruly packet flood into something that behaves like a conversation. It does so with a handful of simple social rules, enforced by mathematics and timers.

First rule: introduce yourself properly. Before two computers exchange a reliable stream of bytes, they perform a small ritual known as the *three-way handshake*. The phrase sounds grand, but the logic is practical. If two parties are to keep track of what was sent and what was received, they need a shared reference point. TCP uses *sequence numbers* for this: each byte in the stream is numbered, and acknowledgments refer to these numbers so the sender knows what the receiver has successfully gotten.

Sequence numbers, however, are not stamped by a universal clock that all machines agree on. Networks contain old packets that can be delayed, duplicated, or reappear out of order. Without a careful start, a new connection might accidentally treat a stray old packet as part of the present conversation. The three-way handshake solves this by having each side announce an initial sequence number and confirm the other's announcement. In simplified terms:

- One machine says, "I would like to talk, and here is where my numbering will begin" (a message with the SYN flag).
- The other replies, "I hear you, and here is where my numbering will begin" (SYN plus ACK).

- The first confirms, “I hear you too” (ACK).

It is not just ceremony. It is the moment the two machines synchronize their expectations, establishing that each can hear the other *now*, not last week, not a millisecond ago on a different path, but in the present connection.

Second rule: acknowledge what you receive. TCP’s ACKs (acknowledgments) are the small nods of digital conversation. An ACK does not say, “I loved your point.” It says, “I have received up to byte number N, and I am ready for the next.” If acknowledgments do not arrive, the sender assumes something went missing and retransmits. This is how the internet becomes reliable without requiring the underlying network to be reliable. Reliability is not in the road; it is in the manners of the travelers.

Third rule: do not speak faster than the other side can listen. TCP includes *flow control*, which prevents a fast sender from overwhelming a slow receiver. The receiver advertises a window: how much data it can accept before its buffers fill. The sender respects this window, pausing or slowing as needed. In human terms, it is the difference between a conversation and a lecture shouted at someone who cannot take notes fast enough.

Fourth rule: do not clog the room. This is *congestion control*, one of the most consequential pieces of network etiquette ever invented. Congestion is not the same as a slow receiver; it is the network itself becoming overloaded, with routers’ queues filling and packets being dropped. If every sender tried to retransmit aggressively the moment a packet was lost, the network would spiral into collapse: loss would cause more traffic, which would cause more loss, which would cause more retransmission, a feedback loop of panic.

TCP’s congestion control teaches endpoints to infer congestion and back off, often by gradually increasing sending rates until loss occurs, then reducing speed and probing again. The details evolved over decades, but the underlying social contract is stable: when the network shows signs of distress, responsible participants slow down. The internet works at scale not because every link is fast, but because the endpoints practice restraint.

These rules created a remarkable outcome: a global network where the intermediate devices (routers) can remain comparatively simple. Routers do not need to know about your file transfer, your video call, or your login. They do not need to reserve a circuit or maintain per-conversation state in the way old telephone switches did. They forward packets based on destination addresses and local routing tables. The intelligence needed for reliability, ordering, and fairness lives at the edges, in the endpoints. This design philosophy has been called the end-to-end principle, but even without the slogan, the intuition is straightforward: if the network is shared by everyone, the middle should not become a fragile bureaucracy.

The everyday experience of the web hides these negotiations because they happen in fractions of a second. But they are there, and they can be felt when something goes wrong.

Imagine clicking a link on a congested café Wi-Fi. The page begins to load, then stalls. Behind the scenes, packets are being dropped, perhaps because too many customers are uploading photos at once, perhaps because interference from neighboring networks corrupts frames, perhaps because the access point's buffers are bloated. Your computer notices missing acknowledgments. It retransmits. It reduces its sending rate, politely stepping back from the crowd. The page trickles in. No one hears a busy signal. No one is told, "All circuits are occupied." Instead, the connection degrades gracefully. The system does not promise steadiness; it promises persistence.

That difference between refusal and degradation is one of the internet's cultural fingerprints. Circuit-switched networks tend to fail with a clear boundary: you are either connected, or you are not. Packet-switched networks tend to fail with ambiguity: you are connected in principle, but performance varies wildly. This ambiguity can be maddening, but it is also a form of inclusivity. Many more people can share the network, many more applications can coexist, because the system does not require exclusive reservation to begin speaking.

The analogy becomes clearer when you compare a dedicated pipe to a city's road network. A dedicated pipe works wonderfully if you can afford to build one pipe per pair of houses and guarantee no interruptions. A road network works because cars can share lanes, detour around construction, and choose different routes. You accept traffic jams in exchange for not needing to build a private highway for every trip.

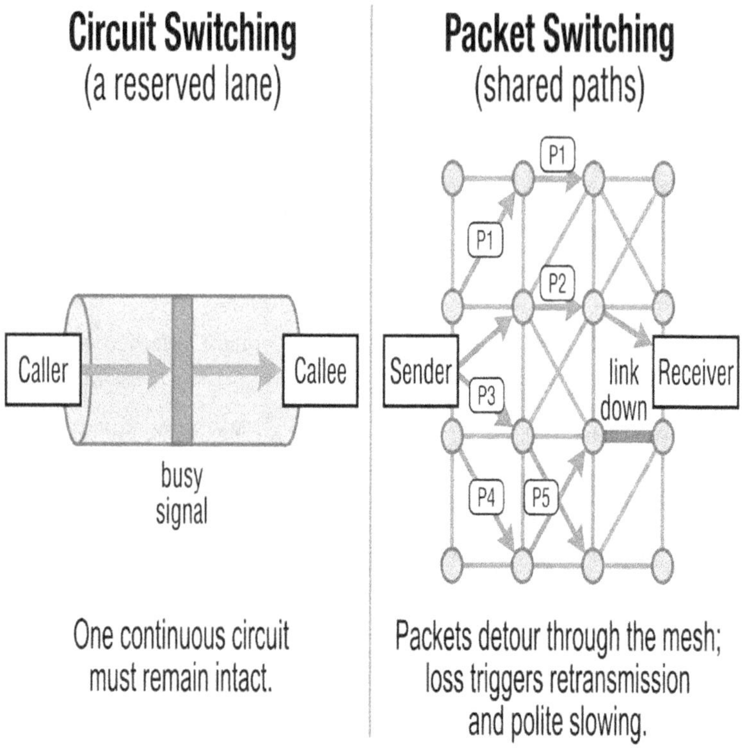

The busy signal on the left side of the diagram is not merely nostalgia; it is a design philosophy. A circuit-switched system rejects a call when it cannot guarantee the exclusive path. A packet-switched system tends to accept the traffic and then struggle in public. This public struggling is precisely what TCP's manners are meant to manage.

A concrete example makes the stakes clearer. Suppose you send a photograph from your phone to a friend overseas. The photograph may be several megabytes. It will be broken into thousands of packets. Those packets will traverse a home Wi-Fi network, then a local ISP, then a regional network, then one or more backbone networks, perhaps an undersea cable, then another ISP, then a mobile network, and finally your friend's phone. Along the way, some packets will be delayed behind others, some will take a route that is momentarily longer, and some may be dropped and resent. Yet the application on your phone expects the photograph to arrive as an intact file. It does not want to hear about the drama of the journey. TCP's job is to absorb that drama and deliver a clean stream of bytes to the application, as if there were a dedicated pipe after all.

The remarkable part is not that this works when the network is quiet. The remarkable part is that it mostly works when the network is full of competing intentions. In a single minute, the same path may carry streaming video, voice calls, software updates, and the routine chatter that keeps cloud services synchronized. Packet switching is not only a technical approach; it is a political arrangement. It assumes that a shared infrastructure can be managed through rules and feedback rather than through rigid reservations.

The shift also changed what it means to *pay* for communication. Circuit-switched billing matched the physics: a call tied up a circuit, so you paid for time. Packet-switched networks decouple cost from duration. A silent TCP connection consumes little capacity; a bursty application can send a lot quickly and then pause. This is one reason internet pricing has always been contentious. People still reach instinctively for time-based metaphors, but packet networks are not renting you a corridor; they are admitting you into a crowd.

Of course, politeness has limits. TCP can retransmit missing data, but if losses are extreme, retransmission becomes its own burden. TCP can slow down for congestion, but if buffers are poorly managed, delay can grow until the connection feels broken even though it is technically alive. And not every application wants perfect reliability. Real-time voice and video often prefer timely delivery over perfect delivery; a late packet is useless. Many such systems use UDP, a simpler transport that provides less etiquette and more freedom.

Even then, the underlying packet-switched logic remains: packets are thrown into the shared network, and the application decides what to do when some do not arrive.

Still, for most of the internet's daily work—loading web pages, sending email, syncing files—TCP's polite conversation is the reason the experience feels coherent. It takes a chaotic substrate and coaxes it into something like orderly speech.

There is a subtle emotional consequence to this architecture. Circuit switching offered a kind of assurance: when the call connected, you felt the world lock into place. Packet switching offers a different reassurance: not that the path is reserved, but that the system can recover. It is the reassurance of redundancy, of second chances, of messages that can be resent, rerouted, and pieced back together. It resembles the resilience of cities rather than the solidity of pipelines: a million small adjustments, constantly negotiated, mostly invisible, keeping the whole arrangement functional.

A human conversation survives interruptions because the participants cooperate. The internet survives interruptions because its protocols assume cooperation, measure it, and enforce it. A packet goes missing; an acknowledgment fails to arrive; the sender tries again, perhaps more slowly, perhaps along a different path. The tone is not the sharp refusal of the busy signal. It is the steady, impersonal patience of a system that expects the world to be imperfect, and continues to talk anyway.

2.2 The Phonebook of Babel

A web address looks like a sentence fragment written for humans: *example.com*, *weather.gov*, *yourbank.com*. It carries intention. It hints at trust. It is memorable in the way a street name is memorable.

The internet itself cannot use that kind of language. Packets do not travel to “the bank”. Routers forward traffic toward numerical destinations: IP addresses, strings like 203.0.113.7 or $2001 : db8 :: 1$. Those numbers are not designed to be lovable. They are designed to fit into routing tables and to be aggregated, advertised, withdrawn, and rewritten as networks grow and reorganize.

The Domain Name System, DNS, is the translator between those two worlds. It is often described as a phonebook, but that metaphor is only half true. A phonebook is a published artifact: one authoritative list, printed once, distributed widely, and corrected slowly. DNS is a living bureaucracy, spread across the globe, built out of agreements about who is allowed to name what. It is less like a book and more like a federation of clerks, each responsible for a drawer of index cards, each with the authority to say, “for names in my drawer, this is the answer.”

This matters because the internet is not only a technical system; it is a system of jurisdiction. In a packet-switched world, you can send a packet almost anywhere if you know the numerical address. DNS decides how easily you can find that address, how quickly it can be changed, and who gets to say what a name means. It turns the casual act of typing a URL into a chain of delegated trust.

A historical detail gives the modern system its edge. In the early days of ARPANET, there really was a single shared file: a list of hostnames and their corresponding addresses, maintained centrally at the Stanford Research Institute. If you wanted to add a new computer, you asked for a new entry. As the network grew, this arrangement turned into a bottleneck and a political headache. Names are not just labels; they are claims. People argue about them, try to squat on them, imitate them, and defend them. A single centralized list does not merely become slow; it becomes contentious.

Paul Mockapetris's design in 1983—1987, captured in the foundational DNS specifications, solved the scaling problem by giving up the fantasy of one master list. The key move was to make naming *hierarchical* and to make authority *delegable*. Instead of asking one office to maintain every entry, DNS builds a tree of names and allows control to be handed down along the branches.

The tree is visible in the punctuation of a domain name, read from right to left. In *www.example.com*, the far right label, *com*, is a top-level domain (a TLD). To its left is *example*, a domain registered under *com*. To its left is *www*, a name under *example.com*. Each dot is not merely a separator; it is a seam where authority can be cut and handed off.

At the very top sits a peculiar character: the root, written as a trailing dot that most people never type. Fully spelled out, *www.example.com.* ends with a dot that means, "this name is absolute, and above it there is nothing." The root is the one point in the DNS tree that does not have a parent. It is not a server, but a small, carefully managed set of information called the root zone: a directory of where the authoritative servers for each TLD can be found. From that directory flow many practical consequences: which organizations operate which TLDs, how changes propagate, and what happens when politics collides with packets.

The authority structure is easiest to understand by watching a single lookup. Imagine typing *www.nytimes.com* into a browser. Your computer does not generally ask the root servers directly. It asks a *resolver*, usually provided by your ISP, your employer, or a public DNS service. The resolver acts like a librarian who knows how to query the system efficiently and who remembers answers for a while so the next person does not need to repeat the whole process.

If the resolver does not already have the answer cached, it begins at the top. It asks a root server: "Who is authoritative for *.com*?" The root server replies not with the address of *www.nytimes.com*, but with a referral: "For *.com*, ask these nameservers." The resolver then asks one of the *.com* TLD servers: "Who is authoritative for *nytimes.com*?" Again, it receives a referral: "For *nytimes.com*, ask these nameservers." Finally, it asks one of the authoritative nameservers for *nytimes.com*: "What is the address for *www.nytimes.com*?" That server replies with one or more records containing IP addresses. Only then can the browser begin the familiar TCP handshake and the rest of the polite conversation that makes the web feel reliable.

The system is simple to describe and maddening to operate. It is distributed, but not flat. The root delegates to TLDs; TLDs delegate to domains; domains may delegate further to subdomains. Each delegation is a boundary of responsibility, and DNS has a formal term for the chunks of the tree that a given set of servers is responsible for: *zones*. A zone is not the same as a domain name in the everyday sense. It is a portion of the namespace for which a server can answer authoritatively. The difference becomes important the moment a large organization subdivides control. A university might control *university.edu* but delegate *cs.university.edu* to a departmental IT team. The departmental nameservers become authoritative for that subtree. The university keeps authority over the rest.

The word *authoritative* is the heart of the matter. DNS responses can be marked to indicate whether the server is speaking from its own authority or repeating what it has heard elsewhere. An authoritative server is the original clerk for that drawer of index cards. A caching resolver is more like a helpful friend who remembers what the clerk said last time. That memory saves enormous time. Without caching, the DNS would creak under its own weight, because every web page load can trigger multiple name lookups, not just for the main site but for images, scripts, tracking pixels, content delivery networks, and advertising partners.

Caching introduces its own trade-offs. DNS records come with an expiration time, a *time to live* (TTL), that tells caches how long they may reuse the answer. A long TTL makes a service faster and reduces load on authoritative servers, but it slows down changes. A short TTL allows rapid failover during outages and makes load balancing more responsive, but it increases query volume and can expose more user activity to resolvers. A DNS administrator choosing a TTL is not merely tuning performance; they are choosing a kind of social contract about stability and agility.

A concrete example brings this to life. Many large websites do not have one fixed server address. They use content delivery networks (CDNs) that place servers near users around the world. DNS becomes part of the steering mechanism. When you ask for *video.example.com*, the answer you receive in New York may differ from the answer you receive in Los Angeles, because the authoritative system (often with the help of the resolver's location) tries to direct you to a nearby cache. DNS looks like a static dictionary, but it is often a dynamic traffic signal.

This dynamism can be a source of both resilience and power. If a datacenter fails, DNS can be updated to point users elsewhere. If a service is under attack, DNS can be manipulated to sink traffic into protective scrubbing centers. But the same mechanism can be used for coercion. If you control a resolver, you can decide what names mean for your users. You can block a domain by returning no answer. You can redirect it by returning the wrong one. In some countries, DNS filtering is a blunt instrument of censorship. In corporate settings, it can be a blunt instrument of security policy. In both cases, the point is the same: DNS is a lever because it sits at the moment where intention becomes an address.

The design of DNS made one controversial decision early on: it prioritized availability and scalability over cryptographic certainty. Classic DNS answers are not signed by default. A resolver receives a response and, unless additional measures are taken, has no built-in proof that the response is genuine. In the 1990s and early 2000s, this was treated as an acceptable risk. The network community was smaller, and the threat model was less industrialized. Then a series of attacks made the weakness impossible to ignore, culminating in widely publicized demonstrations that attackers could poison caches: trick resolvers into storing a false answer so that users would be silently redirected to malicious servers.

The response was DNSSEC, the Domain Name System Security Extensions, standardized in the early 2000s. DNSSEC adds a layer of cryptographic signatures to DNS data. The core idea is simple: authoritative zones publish public keys (in *DNSKEY* records) and use corresponding private keys to sign sets of records (producing *RRSIG* signatures). A resolver that validates DNSSEC can check those signatures and gain confidence that the data was not altered in transit.

The hard part is not the mathematics; it is the chain of authority. DNSSEC needs a way to connect a child zone's key to its parent's authority, or else anyone could publish a signed lie under an impostor key. The mechanism is a record called *DS* (delegation signer), placed in the parent zone to vouch for the child's key. In effect, the parent says, "for this delegated child, here is the fingerprint of the key you should trust." Validation then becomes a walk down the tree, beginning from a trusted starting point, a *trust anchor* (typically the root zone's key). The result is elegant: a cryptographic version of the same delegation structure that already governs naming.

DNSSEC also solves a subtler problem: how to prove that something *does not exist*. Without special care, an attacker could respond to a query for a nonexistent name with a convincing lie, and the resolver would have no way to tell whether the name truly lacked a record or whether the answer was forged. DNSSEC introduces records such as *NSEC* to provide authenticated denial of existence. It is the equivalent of a clerk saying, with a signed statement, "I have checked my drawer, and there is no entry between these two alphabetized cards." This protects against certain kinds of forgery, although it introduces its own privacy complications by revealing structure about zones.

A crucial limitation remains: DNSSEC does not encrypt DNS. It can provide integrity and authenticity, but not confidentiality. Even with DNSSEC, a passive observer watching the network can often see which names you are looking up, because traditional DNS queries and responses are sent in plaintext. This is one reason encrypted transports for DNS have become popular (DNS over TLS, DNS over HTTPS, and related approaches). They change who can see your DNS traffic, and they can improve privacy against local eavesdroppers, but they also shift power toward whoever operates the encrypted resolver, because that operator now sees the user's queries more reliably. The privacy problem does not disappear; it changes shape.

The DNS hierarchy therefore has a double character. Technically, it is a distributed database optimized for answering simple questions quickly: "What is the address for this name?" Politically, it is a nested set of delegations, each one a locus of control. A TLD operator controls policies about registrations and nameservers. A registrar mediates who gets a name and under what terms. A domain owner controls the authoritative zone for that name. A subdomain operator may control their slice. And at the user's end, the resolver can be an invisible governor, quietly deciding which answers are returned and which are withheld.

Even the most mundane DNS record types reveal this layered power. The *A* record maps a name to an IPv4 address; *AAAA* does the same for IPv6. *CNAME* creates an alias, allowing one name to point to another, which is often how services stitch together complex infrastructures while presenting a simple public face. *NS* records name the authoritative servers for a zone, and these are the bones of delegation itself. Each record is a small fact, but the system built from them is a living map of organizational boundaries.

Recent additions to DNS make the map even more operational. Records such as *SVCB* and *HTTPS* allow a domain to advertise service parameters: which protocols it supports, which alternative endpoints exist, and other hints that can help clients connect more efficiently and, in some designs, more privately. This is DNS doing more than naming; it is DNS helping to shape the connection that follows naming. The temptation is obvious: once you have a fast, cached, globally distributed directory, you want to use it for more than one kind of lookup. DNS was designed to be extensible, and the internet has taken that invitation seriously.

The same extensibility that makes DNS powerful also makes it a site of constant negotiation. Every new record type must coexist with old software, middleboxes, and assumptions. Every new use of DNS data is an opportunity for unintended leakage: a record meant to improve performance can increase fingerprinting, revealing something about a client's capabilities. A change meant to improve privacy can interfere with debugging or enterprise monitoring. The hidden internet is full of these trade-offs, where the same mechanism serves multiple masters.

The most vivid way to appreciate DNS is to notice its absence. When DNS fails, the internet can feel as if it has vanished, even though the underlying packet network is still carrying traffic perfectly well. You can often still reach known IP addresses. You can still ping machines. But most users do not navigate by IP number any more than they navigate a city by latitude and longitude. Names are the interface between human memory and machine routing. Break that interface and the network becomes a labyrinth of numbers.

Calling it the Phonebook of Babel is therefore not just a flourish. Babel was a story about language and power: about a shared project collapsing into mutual incomprehension. DNS prevents a kind of daily Babel by giving the world a shared naming grammar, but it does so through hierarchy, delegation, and rules about who may speak with authority. It is a system that turns the right to answer a question into a form of governance.

In the end, a DNS lookup is an act of trust disguised as a convenience. You ask, "where is this?" and you accept an answer from a chain of authorities you will never meet, delivered by caches you do not control, shaped by policies you did not write. Most of the time, the answer is correct, and the world feels smooth. The address resolves, the packets flow, and a human name becomes a numerical destination, as if language itself had been routed successfully across the planet.

2.3 Whispers in the Dark

The moment of unease usually arrives quietly. You are on airport WiFi, or in a hotel, or on a train where the network name sounds plausible but not familiar. Your phone connects anyway, because convenience is what networks sell. Then you open a banking app, type a password, and feel a small hesitation: who else can see this?

On the early internet, that hesitation had an obvious answer. Anyone on the same local network could often read traffic as easily as overhearing a conversation in the next seat. A packet-switched network is a shared medium by design; packets pass through routers, switches, access points, and sometimes through devices owned by organizations you have never heard of. Even if those intermediaries are honest, the network environment is noisy and crowded, and the simplest version of the internet makes no promise of privacy. The polite conversation of TCP keeps packets in order and retransmits what is lost, but it does not make them secret.

Secrecy requires a different kind of agreement: not about *delivery*, but about *trust*. The most widely used solution is TLS, Transport Layer Security, which wraps ordinary connections in encryption and authentication. TLS is the reason a browser shows a padlock for an *https://* site, and it is the reason an eavesdropper on that airport WiFi cannot simply read your password as it passes.

TLS does something that sounds impossible at first: it allows two strangers to create a shared secret in public.

Two problems must be solved at once. First, the data must be encrypted so that observers cannot read it. Second, you need confidence about whom you are talking to. If an attacker can impersonate your bank, you can end up whispering your secrets to the wrong person with perfect privacy *from everyone else*.

A good way to picture the underlying magic is an old metaphor: the open padlock. Imagine Bob wants anyone to be able to send him a locked box, but only Bob should be able to open it. He buys a special kind of padlock that can be snapped shut but cannot be opened without its key. He keeps the key in his pocket, and he hands out copies of the *open* padlock to the public. Alice, wanting to send Bob a secret letter, puts the letter in a box, closes Bob's open padlock, and sends the locked box across the world. Anyone can intercept it. Anyone can hold it. No one can open it without Bob's key.

That is the intuition of public key cryptography. The open padlock corresponds to a *public key*, a piece of information meant to be shared widely. The key in Bob's pocket corresponds to the *private key*, kept secret. What makes this feel like a paradox is that the public key and private key are mathematically linked. The public key can be used to encrypt or to verify signatures, and the private key can be used to decrypt or to create signatures. The details are complicated, but the effect is practical: it becomes possible to communicate securely without meeting in advance to exchange a secret password.

TLS uses this machinery in a pragmatic way. It does not, in modern designs, encrypt the entire data stream with public key cryptography, because that would be slow. Instead it uses public key techniques to agree on temporary *symmetric* keys: ordinary shared secrets that can encrypt large amounts of data efficiently. Public key cryptography is the ceremony at the door; symmetric encryption is the quiet conversation once you are inside.

This ceremony is called the TLS *handshake*. The word is apt. Two people can shake hands in a crowded room, and the handshake itself announces something: a mutual recognition, a decision to interact. The handshake does not by itself convey the content of the conversation, but it establishes the relationship that makes conversation possible.

In a modern TLS handshake, the client (your browser, your app) begins by sending a message that amounts to: "Here are the cryptographic options I support, and here is a fresh piece of randomness to help create new keys." The server replies with its choice of options and its own contribution to key creation. Crucially, the server also sends a *certificate*, an identity document for the internet.

A certificate is an X.509 document that binds a public key to a name, such as *www.yourbank.com*. It says, in effect, "this public key belongs to whoever controls this domain." The certificate is digitally signed by a certificate authority (CA), an organization that vouches for the binding. Your device carries a pre-installed list of CA public keys—its set of entities it is willing to trust as introducers. This is the Web PKI, the web's public key infrastructure.

If that sounds like a lot of social trust smuggled into a technical mechanism, it is. The cryptography can prove that a certificate was signed by a CA. It cannot prove that the CA was wise to sign it. That is a human governance problem. The internet solves it the way it solves many governance problems: by layering, auditing, and collective pressure, rather than by finding a perfect authority.

This layered trust can feel abstract until you see what happens when it fails. In 2011, an attacker compromised a Dutch certificate authority called DigiNotar and obtained fraudulent certificates for domains including Google. Those certificates were reportedly used in the wild, enabling man-in-the-middle attacks in which victims could be silently intercepted while seeing what looked like a secure connection. The cryptography still worked. The signatures still verified. The problem was that the wrong entity had been trusted to vouch for identity. Browser makers responded by revoking trust in DigiNotar, an extraordinary event: a globally distributed software ecosystem collectively decided that one CA was no longer an acceptable introducer.

The DigiNotar incident is a reminder that "trust" in TLS is not primarily about believing in mathematics. It is about deciding which organizations are allowed to make identity claims on behalf of the rest of us. The mathematics enforces the consequences of that decision.

TLS 1.3, standardized in 2018, tightened the ceremony. One important change is that much of the handshake becomes encrypted earlier in the process, reducing what passive observers can learn. The principle is straightforward: if the handshake is the act of agreeing on a secret, then the details of that agreement should not be fully visible to bystanders. TLS 1.3 also strengthened protections against downgrade attacks, where an attacker tries to trick two parties into using an older, weaker version of the protocol.

Even with these improvements, the handshake inevitably reveals some information. An on-path observer can still see the source and destination IP addresses, and can often infer the general type of traffic. But a subtler leak has had outsized consequences: the SNI.

SNI stands for Server Name Indication. It exists because the web evolved in an economical direction: *virtual hosting*. One physical server, or one IP address, often hosts many different websites. When a client connects to that IP address, the server needs to know which site the client wants in order to present the correct certificate. Early TLS did not include a way for the client to say, "I'm trying to reach *this* hostname." SNI, introduced as an extension, fixed that by allowing the client to include the target hostname in its initial hello message.

The fix came with a privacy cost: classic SNI is sent in plaintext. Someone watching the network might not be able to read the content of your connection, but they can often see the hostname you are connecting to. For many people, that is already sensitive. Knowing that someone is visiting a particular medical site, labor union site, religious site, or news outlet can be enough to profile them, even without reading the page.

This is not only about government surveillance. It is also about commercial monitoring, workplace oversight, and the casual curiosity of anyone operating a network you happen to use. Encryption that hides content but leaves destinations legible is like whispering while pointing.

The response has been another layer of etiquette: Encrypted ClientHello (ECH), an evolving standard intended to hide SNI and related handshake information by encrypting parts of the client's initial message. The idea is again paradoxical: encrypt the greeting before you have finished agreeing on keys. ECH does this by giving the client a public key for the server's *front door*, allowing the client to encrypt the sensitive parts of its hello so that only the intended server can read them.

Here the layers begin to intertwine. How does the client learn the key needed for ECH? In many designs, it is published through DNS, using extensible records such as *HTTPS* or *SVCB* records. The naming system, already the translator between intention and address, becomes a distributor of cryptographic hints about how to connect privately. This is a recurring pattern in the hidden internet: one layer, designed for a narrow job, becomes the messenger for another layer's secrets.

A concrete example helps. Suppose you type *www.example.com* into a browser. A DNS lookup returns an address. But increasingly, DNS may also return additional information: perhaps that the site supports HTTP/3, perhaps that it prefers encrypted DNS, perhaps that it offers ECH keys. Before any webpage loads, a quiet negotiation has already occurred. The browser is not merely asking "where is it?" but also "how should I speak to it?"

All of this complexity exists for a simple emotional goal: the ability to act online without feeling watched. Yet it is worth being honest about what TLS can and cannot do.

TLS can provide confidentiality of the data between your device and the server you reach. It can provide integrity, ensuring data is not altered in transit. It can provide server authentication, so your browser can reject impostors *if* the certificate system is functioning properly and *if* you pay attention to warnings.

TLS cannot hide everything. The IP address you connect to remains visible to networks along the path, and large services often have distinctive address ranges. Traffic patterns can leak information: timing, packet sizes, and connection frequency can act like a fingerprint. If DNS queries are unencrypted, those queries can reveal your destinations even if TLS is perfect. Even when DNS is encrypted and ECH is used, other identifiers can betray you, from cookies to device characteristics. TLS is a powerful curtain, but it is not a magic cloak.

Still, it changed the character of the internet. Early web traffic was often sent in plaintext, with security reserved for logins or payments. Today, encryption is closer to the default. This shift has consequences beyond individual privacy. It changes what network operators can inspect and modify. It complicates certain forms of censorship and surveillance while enabling others. It forces security tools to evolve, sometimes moving inspection onto endpoints rather than the network. It reshapes the balance of power between users, ISPs, governments, and platforms.

It also changes how trust feels. In the physical world, trust is often personal: you recognize a face, you know a voice, you remember a history. Online, trust is mediated by chains of certificates and lists of trusted authorities embedded in software. You do not know the people running the systems you rely on, and often they do not know you. The relationship is formal, contractual, and to some degree automated. When it works, it is liberating: you can buy a book from a small shop across the world, or send a confidential message, without asking permission from anyone along the route. When it fails, it is uncanny: you can be redirected by a forged certificate, or tricked by a lookalike domain, and the familiar signals of safety may still appear.

Perhaps the most revealing thing about TLS is how much effort goes into protecting the handshake itself. The content of a conversation matters, but so does the fact *that* a conversation is happening, *with whom*, and *under what pretense*. Privacy is not only about hiding words; it is about controlling what relationships are legible to strangers. On a shared, packet-switched network, the ability to whisper is not a luxury feature. It is the difference between living in a city where anyone can read your mail and living in one where envelopes are assumed.

A secure connection, at its best, is a small triumph of cooperation among strangers: protocol designers who anticipated failure, engineers who maintained keys and certificates, browser vendors who enforced rules, and users who clicked a link without needing to become cryptographers. The packets still cross networks owned by others. They still traverse cables and routers and crowded airwaves. But wrapped in encryption, they become less like postcards and more like sealed letters, carried through the same public world with their contents reserved for the intended hands.

Chapter 3

The Bazaar at the Edge of the World

We tend to view the internet as a monolithic cloud, but it is actually a fractious confederation of thousands of independent networks trading traffic like commodities. This chapter descends into the messy reality of routing tables and exchange points, revealing how the global flow of information relies not on central command, but on a fragile web of trust and handshake deals between corporate rivals.

3.1 The Cartography of Chaos

If the Internet had been designed like a national highway system, there would be a central authority that assigns routes, publishes a definitive atlas, and updates it when bridges collapse. Instead, the Internet grows the way port cities grow: around trade, rivalry, accidents of geography, and deals made at odd hours by people who are trying to keep things running. The protocol that makes this possible is called the Border Gateway Protocol, or BGP. It is not a map of the Internet so much as a way of exchanging *claims* about how to reach places, and then choosing among those claims according to local preference.

That word *border* is more political than technical. Inside a network operated by a single organization—a university, a cloud provider, a national ISP—routing can be run like a private road system. At the edges, where one operator's roads meet another's, there is no shared boss. BGP is the language used at those edges. It allows independent networks to tell one another, in a standardized way, "I can deliver packets to these destinations, and here is the sequence of networks you would pass through if you send them to me."

A crucial implication follows: there is no single "the Internet" in the sense of a single, authoritative diagram. There are many overlapping views, each shaped by where you stand and by the deals your network has made. Two engineers in different companies can look at the same destination and see different best routes, both correct *for them.* Even within one company, a router in New York may make a different choice than a router in Los Angeles, because the options and costs differ. Cartography becomes local, contingent, and continuously revised.

A useful unit of citizenship in this world is the *Autonomous System* (AS). An AS is a collection of IP networks and routers under one administrative control, presenting a common routing policy to the outside world. The key phrase is *routing policy*. An AS is less a blob of cables than a promise: a promise that, within its borders, it can make internal decisions privately, and at its borders, it can speak BGP in a coherent voice.

Each AS is identified by a number. Some of those numbers are famous in the industry the way airline codes are famous in airports: you may not know what they mean, but you keep bumping into them. When BGP announces reachability, it does so in terms of *prefixes*—blocks of IP addresses, like a postal code range. A network might announce, for example, that it can reach 203.0.113.0/24. That "/24" is shorthand for how large the block is. The announcement does not say where those machines physically sit. It says only: *if you give me packets for this prefix, I know what to do with them.*

The announcement also carries a kind of provenance: a list of AS numbers called the *AS path*. This is BGP's signature move. Rather than distribute a complete map and compute routes from it, BGP distributes route advertisements that carry their own travel history. When an AS passes along a route, it adds itself to the front of the list. The route becomes a chain of custody: to reach a given prefix, the packet would go through AS 64510, then AS 64520, then AS 64530, and so on.

This list is the reason BGP is often described as a *path-vector* protocol. It is also the reason the Internet usually avoids certain kinds of catastrophic loops. If an AS sees its own number already in the AS path, it knows that accepting the route would create a cycle. That simple check has kept countless misconfigurations from turning into global whirlpools.

The AS path, however, is not a lie detector. It is a statement made by a neighbor, and neighbors can be wrong. They can be mistaken, or sloppy, or malicious, or merely trapped in a subtle situation where two policies collide. BGP was built for a community that largely trusted one another, a community where the main problem was how to interconnect heterogeneous networks, not how to defend against deception at planetary scale.

A common misunderstanding is to imagine that Internet routing is mainly about finding the *shortest* path. That is how many people first learn about networks: routers as diligent mathematicians, computing the minimal number of hops or the minimal delay. BGP is not that. BGP's job is to choose a *permitted* and *preferred* path given business relationships and operator choices, and only then to use tiebreakers that resemble technical optimization.

Inside a single organization, shortest paths can be computed against a shared view of the topology. Between organizations, the topology itself is part of the competition. The fact that you could, in principle, deliver a packet over a certain sequence of networks does not mean those networks will agree to carry it, or agree to carry it for free, or agree to carry it in both directions. Much of the Internet's shape is carved by contracts: customer relationships, peering arrangements, regional restrictions, and obligations to keep traffic within certain jurisdictions.

So BGP behaves like a bazaar rather than a railway timetable. Each AS runs its own selection process, guided by local policy. "Prefer routes learned from customers." "Do not send this kind of traffic through that neighbor." "If two routes are otherwise equal, prefer the one that exits my network closest to the destination." These are not standardized moral laws; they are configuration choices, tuned to money, performance, and risk.

The result is a kind of organized chaos: a global system that works because many local optimizations and partial truths happen to align most of the time.

A concrete picture helps. Imagine three Autonomous Systems:

- AS 64530 owns the destination prefix 203.0.113.0/24 (perhaps a service hosted in its data centre).
- AS 64520 sits in the middle and offers connectivity onward.
- AS 64510 is an access provider with end-users trying to reach that prefix.

Physically, perhaps AS 64510 has a direct cross-connect to AS 64530 in the same building, while AS 64520 is across town. A human looking at a map would say: *go directly; it is shorter.* But economics can flip the choice. If the direct link requires a paid arrangement that AS 64510 does not have, while AS 64520 offers cheap transit as part of a bundle, then the permitted path may be the longer one. BGP's job is to make that permitted path visible and selectable.

Path Vector Logic

AS_PATH grows as the update propagates.

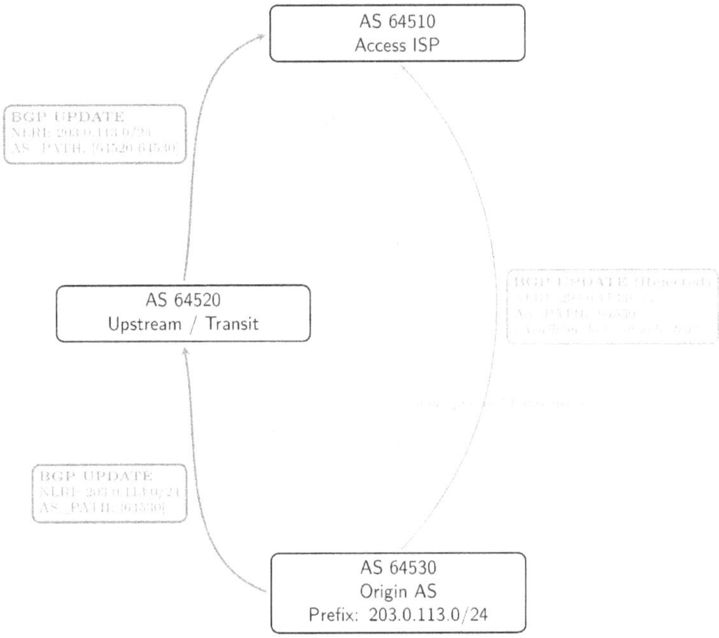

Policy vs Physics

The chosen route (Blue) is longer in distance because business policy overrides the shortest cable path.

The diagram is deliberately abstract. In reality, each of these lines might represent multiple routers, redundant links, and dozens of alternative paths. But the essential idea survives scaling: BGP is a system of announcements that accumulate a trail of AS numbers, and the route that wins is not necessarily the one a geographer would pick.

Once you see BGP as a rumor network, many oddities of everyday Internet life become easier to interpret. Why does a website hosted "nearby" sometimes feel far away? Because *nearby* in kilometres may be *expensive* in contracts. Why does traffic between two cities in the same country sometimes detour through another country? Because the available interconnection points and commercial agreements may make that detour the permitted path. The Internet can be fast and still be circuitous.

BGP also explains a deeper, more unsettling fact: the Internet's global behavior depends on a vast amount of human-maintained configuration. Each AS is constantly making choices about which routes to accept from neighbors, which to advertise onward, and which to prefer. These choices are encoded in router configurations that can be tens of thousands of lines long, influenced by corporate policies, regulatory obligations, and the practical scars of previous outages.

This is where the fragility enters. A single incorrect announcement can leak out of one AS and be accepted by others, not because they are foolish, but because the system is built to accept neighbor claims quickly. The Internet values reachability and fast convergence. When a path disappears, BGP tries to find another. When a new route appears, BGP tries to use it. That responsiveness is part of its resilience. It is also how mistakes propagate.

A historical episode made this fragility vivid for many engineers: the YouTube hijack of February 2008. Pakistan Telecom attempted to block YouTube domestically by advertising a route to YouTube's address space that directed traffic into a sinkhole. The intention was local: make Pakistani users unable to reach YouTube. But the announcement escaped. Other networks accepted the more specific route and propagated it. For a time, significant portions of the world tried to reach YouTube by sending traffic toward Pakistan Telecom, where it was blackholed. The Internet did what it is designed to do: it believed a route announcement and converged on it. The fact that the announcement was meant as censorship rather than connectivity was, to BGP, irrelevant.

The episode was not the first routing leak, and it would not be the last, but it became a frequently cited parable because it combined a political motive with a technical mechanism that had no built-in notion of political boundaries. A tool used to enforce a national policy became, briefly, a lever on global reachability.

It also exposed an uncomfortable asymmetry: it is easier to make an incorrect route widely visible than it is to retract its effects quickly. Even when the bad announcement is withdrawn, caches, timers, and the sheer inertia of a distributed system can leave some parts of the network converging slower than others. Minutes of disruption can be enough to break financial transactions, disrupt emergency services, or cascade into failures elsewhere.

The fact that BGP is partially observable makes cartography even harder. Researchers and operators build "maps" of the Internet using route collectors: systems that peer with many networks and record the BGP updates they see. These collectors are like weather stations. They give valuable data, but only from where they are placed, and only from the neighbors willing to share.

A route collector might sit at an Internet Exchange Point and receive feeds from dozens or hundreds of networks. It can tell you what routes those networks *announce to the collector*. But it cannot tell you everything those networks announce privately to others. Some relationships are hidden behind private interconnects. Some announcements are tailored: a network may show one set of routes to one neighbor and a different set to another. And because BGP is policy-driven, two collectors in different regions can legitimately see different best paths for the same prefix at the same time.

This partial visibility has a philosophical consequence: any global "Internet map" is an inference. Projects that infer AS relationships from observed paths can be extremely useful, but they are still reconstructions from shadows on a wall. The Internet is not lying; it is simply decentralized in a way that resists a single, clean diagram.

It helps to distinguish two kinds of failure. One is *misconfiguration*: an operator accidentally announces routes they did not intend to announce, or fails to filter routes from a customer, or sets a preference incorrectly. The other is *hijacking*: an actor deliberately announces routes to attract traffic, for surveillance, disruption, or fraud. In both cases the mechanism is similar: BGP spreads reachability claims, and other networks decide whether to believe them.

BGP was designed in an era when the Internet's interconnection community was small enough that trust could be social as much as technical. If a neighbor misbehaved, you called them. If they did not answer, you called someone who knew them. That human web still exists, astonishingly, but it now strains under the scale and geopolitical diversity of the modern Internet.

Security improvements have therefore tended to bolt on *ways of checking claims*, rather than replacing BGP outright. One important direction has been to make it possible to validate which AS is authorized to originate a prefix. Even that, however, is only part of the puzzle. BGP's AS path can be manipulated in subtler ways than a simple false origin, and the protocol's flexibility is part of what makes it operationally useful. Operators want to express policy; security mechanisms must coexist with those policies, not flatten them into a single "correct" route.

A more intimate kind of chaos appears in BGP's everyday dynamics: *convergence*. When a link fails, routers withdraw routes that depended on it, and alternative paths may be announced and selected. During the transition, different parts of the Internet may disagree about the best path, and packets can take strange routes, or be dropped. Some of this is visible to users as transient glitches: a video stream stutters, a call drops, a website intermittently times out. The Internet's core is often robust enough that these events feel like momentary weather, not structural instability. But at scale, convergence behavior is a profound form of collective computation, performed by thousands of independent systems, each obeying local rules.

There is a human comfort in believing that networks are engineered like clocks. BGP is engineered more like a market: a set of incentives, conventions, and rules of thumb that usually produce a workable outcome without guaranteeing a globally optimal one. It can route around failures, but it can also route around common sense. It is resilient, but not always graceful.

Seen this way, the Internet's borders are not lines on a map. They are negotiations encoded in configuration: which routes you accept, which you propagate, which you prefer, and which you reject even if they look technically fine. The AS path is the receipt attached to the rumor. It tells you where the rumor has been, not whether it should have been believed.

And yet the system's daily success remains extraordinary. Every time a phone wakes up and finds a route to a distant server, it is relying on millions of such whispered claims being mostly consistent: on a decentralized agreement about how to deliver tiny envelopes to numbered streets that no one authority governs. The Internet keeps its shape not by possessing a perfect map, but by continuously redrawing a provisional one, line by line, deal by deal, failure by failure, in the ceaseless present tense of BGP updates.

3.2 The Handshake Economy

A packet crossing the Internet resembles a letter that changes postal services mid-route. One carrier takes it across a city, hands it to another for the highway, and a third for the last stretch. The letter does not carry cash. The packet does not carry a credit card. Money moves off to the side, through contracts and invoices and informal norms, while the packet simply obeys the routing decisions that those arrangements permit.

BGP provides the language for announcing reachability, but it does not tell anyone what they *ought* to carry. That question is answered by interconnection: the agreements and physical connections that let one network hand traffic to another. The Internet's grand paradox is that its most global function depends on intensely local acts: two companies agreeing to plug in a cable.

Those acts often happen in places that look deliberately ordinary. A hotel-like building with tinted windows. A converted warehouse with an airlock of badge readers. A data centre corridor where the loudest sound is the wind of fans, and the most valuable objects are the dullest-looking: bundles of fiber, trays of copper, the small plastic handles of patch cords. In this setting, words like *peering* and *transit* are not abstractions. They are the difference between a free handshake and a metered toll.

To make sense of those words, it helps to start with a simple idea: every network has customers, and every network has neighbors. Customers pay you to carry their traffic to the rest of the Internet. Neighbors exchange traffic because it is mutually useful. The two relationships overlap in practice, but the distinction is crisp enough to steer the entire shape of global routing.

Transit is the paid service most people intuitively imagine. A network purchases upstream connectivity from a larger network: "Carry my traffic to the wider Internet, and bring back traffic destined for me." The upstream network becomes a kind of wholesaler of reachability. If you are a small ISP in a rural region, you may buy transit from a national carrier. If you are a content company expanding into a new country, you may buy transit from a local provider until you have enough presence to connect more directly.

In transit, money flows one way and packets flow both ways. The seller is agreeing to provide access to *everything*, subject to technical constraints and acceptable-use terms. Transit is attractive because it is simple: one contract, one or a few physical handoffs, and you can reach the entire world.

Peering is different. Two networks connect directly and exchange traffic between their respective customers without payment, or with payment so small and situation-specific that operators still talk about it in the idiom of barter. The classic form is *settlement-free peering*. Each party agrees: "I will carry traffic from you to my customers, and you will carry traffic from me to yours." No one is promising global reachability. No one is promising to carry traffic to third parties. The deal is narrower and, precisely because it is narrower, it can be free.

Even these simple definitions conceal the most human part of the Internet: peering decisions are often negotiations about fairness, leverage, and risk. A network that delivers popular video may insist that its traffic is a benefit to the access provider, because customers subscribe for that content. The access provider may insist the opposite: that the video network should help pay for the capacity upgrades that keep customers happy. Both can be right, depending on which costs and benefits you count. The packet, meanwhile, does not care. It will travel wherever the contract allows.

A surprising amount of Internet behavior follows from one technical fact: BGP routes are selected and exported according to policy. Policy is where business relationships are enforced. A common rule of thumb can be stated plainly: export customer routes widely, but be selective with peer and provider routes.

If you buy transit from an upstream provider, that provider will give you routes to the full Internet. You will send your customers' traffic to the provider, and the provider will deliver it onward. But you generally will not take routes you learned from one transit provider and advertise them to another transit provider, because that would make you a free middleman, carrying traffic you are not being paid to carry. Similarly, if you peer with another network, you typically exchange only your own routes and those of your customers, not the entire table of everything you learned elsewhere. Peering is not a charity; it is a controlled exchange.

These norms are so widespread that many engineers speak of a *valley-free* world: traffic can go uphill from a customer to a provider, optionally cross to a peer, and then go downhill to a customer, but it should not slosh around between providers. The metaphor is imperfect, but it captures the directional bias that contracts impose on routing. The physical fiber might allow many possible paths; the economic gradient prunes most of them.

The consequences are visible in the strange geometry of the Internet. A physically short route can be economically forbidden, while a longer route is encouraged. Data can detour through a city that hosts a key exchange point, or through a carrier hotel where the right contracts exist. The Internet's pathways look less like straight lines and more like what you get when you overlay a map of roads with a map of trade agreements.

Interconnection happens in two basic physical ways. The first is a *private network interconnect*: a direct link between two networks, often a cross-connect inside a data centre. It can be as literal as a patch cable running from one cage to another. It is private in the sense that only those two parties use it. Private interconnects are common when traffic volumes are high or when a network wants tight control over performance and troubleshooting.

The second is an *Internet Exchange Point*, or IXP. An IXP is a meeting place where many networks connect to the same switching fabric and exchange traffic. The IXP itself is not the Internet and does not forward traffic like a transit provider. Think of it as a well-run marketplace square: the square provides the space and rules that make many trades possible; the traders still decide whom to trade with and what to exchange.

At an IXP, each participant brings a router port to the exchange switch. They can then establish BGP sessions with other participants across that shared infrastructure. The IXP reduces the cost of connecting to many networks. Without it, a network that wanted to peer with N other networks would need N separate physical circuits or cross-connects. With an IXP, the physical connectivity becomes one port to the exchange, and the logical connectivity can scale through many BGP sessions.

The most successful IXPs become magnets. If you run a regional ISP, joining a strong local exchange can dramatically reduce your dependence on distant transit providers. Traffic that used to trombone through another city to reach a neighboring network can be exchanged locally, improving latency and reducing cost. If you are a content provider, joining the same exchange lets you get closer to users without negotiating dozens of bespoke private links. An IXP turns the geography of a city into a piece of Internet architecture.

The word *handshake* is apt because peering is as much social as technical. Engineers talk about *peering policies*: published criteria that specify when a network will peer settlement-free. Some are generous: "We peer openly with anyone at these exchanges." Others are selective: minimum traffic levels, minimum number of locations, requirements about routing hygiene, insistence on 24/7 operational contacts. These policies are not only about cost; they are about risk management. A badly run neighbor can leak routes, cause outages, or become a conduit for abuse. A selective policy is often an attempt to restrict peering to parties likely to behave well.

This social dimension produces an odd kind of paperwork infrastructure. Networks need to find one another, know where they are present, and know whom to contact. Much of that coordination happens through shared registries maintained by the community, where networks list their technical details, peering points, and policies. The information is not handed down by a central authority; it is curated, corrected, and occasionally argued over. Even the metadata of the Internet is a negotiation.

Once two networks agree to peer, the handshake becomes electrical. They choose a location, order cross-connects or ports, and configure routers. The routers establish BGP sessions and exchange routes. From then on, the relationship becomes part contract and part operational routine: monitoring, capacity planning, dealing with congestion, and responding to incidents. The most stable interconnections are boring. The most newsworthy are those where boredom breaks.

A concrete example shows how quickly the handshake economy can become a public argument: the interconnection disputes of the streaming era, when video traffic grew large enough to reshape who bore the cost of capacity. A well-known flashpoint in the United States involved Netflix and Comcast in the early-to-mid 2010s, when customers complained of degraded streaming performance and the companies blamed one another over who should upgrade congested interconnection links. The conflict spilled into press coverage and regulatory debates because it touched a nerve: customers felt they were paying the access provider for a reliable Internet, while content companies felt they were already paying their own transit and delivery costs. The technical symptom was congestion at specific interconnection points, but the underlying issue was bargaining power.

From a BGP perspective, the story is blunt. If two networks do not expand capacity where their traffic meets, packets queue, latency rises, throughput collapses, and users experience “the Internet” as slow. Nothing in the packet explains why it is slow. Nothing in the packet can compel either side to upgrade. The physics of fiber and the math of queues are downstream of a contract negotiation.

Economists and network researchers have tried to formalize this kind of bargaining. Doron Zarchy and Amogh Dhamdhere (2016) framed peering as a bargaining problem in which payments could, in principle, be derived from how much each side benefits, rather than from coarse categories like “peer” and “provider.” That academic framing matters because it mirrors what operators already do informally: argue about who benefits, who has alternatives, and who can walk away. The handshake economy has always been a negotiation; it merely became more visible when video made the stakes obvious to ordinary households.

IXPs complicate the story further, in a good way. By creating dense hubs of interconnection, they can reduce the leverage of any single bilateral relationship. If an access network can reach many content providers through local peering at an exchange, it may depend less on paid transit. If a content provider can reach users through many exchanges, it may avoid expensive private interconnects or contentious negotiations. In that sense, IXPs act like civic infrastructure: they lower barriers to trade and increase the number of potential partners.

But IXPs also introduce their own choices. Some exchanges run *route servers*, which allow participants to peer multilaterally: instead of setting up separate BGP sessions with each neighbor, a network can peer with the route server, which redistributes routes according to policies. This makes peering easier, especially for smaller networks. Yet it also concentrates some operational complexity into shared systems, and it can blur the mental model of "who is my neighbor?" when a single BGP session effectively represents dozens or hundreds of peers.

Even without route servers, an IXP is never just a switch. It is an organization with governance, membership rules, pricing, and a culture. Some are run by non-profits with a public-service ethos; others are run commercially. Some cultivate neutrality; others are enmeshed in local politics. The Internet's supposed immateriality rests on these very material institutions.

For someone outside the industry, the most counterintuitive point may be this: *peering is not automatically good.* Direct connection can reduce latency and cost, but it can also create dependencies. If you peer with a neighbor and that neighbor has an outage, a portion of your traffic may fail in a way that is hard to diagnose, because the failure is outside your administrative control. If you peer at a single exchange and that facility has power issues, your carefully optimized local paths may vanish. Networks therefore tend to build redundancy across locations and relationships, mixing peering and transit in ways that resemble diversification in finance.

This is where the bazaar metaphor becomes more than poetry. In a bazaar, you do not rely on one vendor. You cultivate multiple suppliers, you know who is trustworthy, and you keep an eye on prices. Internet operators do the same, except the commodities are reachability and capacity. The currency is money, but also latency, resilience, and reputation.

BGP then becomes the mechanism for expressing those choices at runtime. Routes learned from a peer at a local exchange may be preferred for performance. Routes learned from an upstream provider may be used as a default. Routes learned from a backup provider may be held in reserve. If congestion appears, engineers may adjust preferences or shift traffic—not because the laws of routing demand it, but because the contracts and the economics make one path more tolerable than another.

The handshake economy can feel unsettling because it replaces the comforting idea of a single public Internet with a web of private bargains. Yet it is also one reason the Internet expanded so quickly. A centrally planned network might have required uniform standards of payment and responsibility before it could grow. The Internet allowed growth through bilateral agreements: two networks could connect, exchange traffic, and create value for their users without needing permission from a global authority. Scale emerged from repetition.

That repetition still has a face. Somewhere in a data centre, a technician is labeling cross-connects. Somewhere in a network operations center, an engineer is watching graphs of port utilization and deciding whether to upgrade a link. Somewhere in an inbox, a peering coordinator is negotiating terms that will decide whether traffic flows directly or takes a detour. The public sees a spinning buffer icon; the hidden Internet sees a relationship that needs to be renewed in steel, glass, and the quiet ritual of plugging one network into another.

3.3 The Physics of the Final Mile

A visitor from another century might find it strange that the most difficult part of a global communication system is the last few hundred meters. Under oceans lie cables as thick as a garden hose, capable of carrying terabits per second. Between cities, long-haul fiber runs in buried conduits and along railroad rights-of-way, lit by lasers and amplified in huts that few people ever notice. Data centres trade traffic in buildings that look like anonymous fortresses. By the time the packet reaches your neighborhood, the Internet has already done the heroic part.

Then it slows down.

The slowdown is not always visible as a lower number on a speed test. It is often felt as a hesitation: the slight pause before a web page begins to render, the delay that makes a video call feel like a walkie-talkie, the jitter that turns online games into a lottery. The physics of the final mile is not only about raw bandwidth. It is about latency, about waiting, about sharing, and about infrastructure that was never built for the kind of demand now placed upon it.

The phrase *final mile* is a convenient lie. In dense cities it may be a few meters of fiber to an apartment building. In suburbs it may be a kilometer of coaxial cable draped through poles and taps. In rural areas it can be many miles of copper pairs strung across long distances, or a radio link that competes with trees, weather, and hills. What unites these cases is not distance but economics: the global core is shared by millions of customers, while the connection to any one customer is largely a bespoke expense. Digging trenches, leasing pole space, obtaining permits, negotiating building access, maintaining equipment in storms, sending technicians to fix faults—these costs do not shrink just because the customer is one among many. The final mile is where the Internet becomes a municipal project.

One way to understand the bottleneck is to compare the Internet to a water system. A city can build a huge aqueduct and a massive reservoir, but the flow into each home still depends on the diameter of the local pipe and the pressure at peak demand. Even if the reservoir is full, the pipe to your kitchen can be narrow. Even if the pipe is wide, the pressure can drop if everyone runs their shower at once. The final mile is that kitchen pipe, and its limitations show up as contention.

Contention is not a scandal; it is the business model. Access networks are built on *oversubscription*: the assumption that not everyone will max out their connection at the same time. If an ISP had to provision dedicated peak capacity for every household simultaneously, consumer broadband would cost far more than most people would pay. Instead, capacity is pooled. When the pool is large enough, the statistical peaks smooth out. When the pool is stressed—during evening video streaming, during a neighborhood event, during a work-from-home surge—queues form, and people experience the Internet as sluggish.

The global core can also be oversubscribed, but it is easier to add capacity there. Fiber in a long-haul conduit can often be upgraded by lighting additional wavelengths, replacing optics, or adding parallel routes where rights-of-way already exist. In the final mile, upgrades tend to be physical: pulling new cable through ducts, replacing amplifiers along streets, rewiring apartment buildings, upgrading headends, and installing new customer equipment. Every upgrade requires labor, access, and coordination. Physics meets bureaucracy.

A second idea matters even more for how the Internet *feels*: latency. Latency is the time it takes for a signal to travel and for the system to respond. In a fiber strand, signals move at a large fraction of the speed of light, but not at the speed of light in vacuum. Over long distances, that propagation delay is unavoidable. Yet the delays that irritate users most are often local: tens of milliseconds added by buffering, scheduling, Wi-Fi contention, or overloaded equipment in the access network.

There are different kinds of delay, and they stack.

Propagation delay is the clean, physics-textbook kind: distance divided by speed. If you are sending packets across an ocean, this dominates, and no amount of cleverness makes the ocean smaller.

Transmission delay is the time to push bits onto a link. On a slow link, large packets take longer to serialize.

Processing delay is the time routers and modems take to examine headers, apply policies, and forward.

Queuing delay is the most treacherous: the waiting time when packets arrive faster than they can be sent. Queuing delay is not constant. It grows with load, and it is often the reason a connection feels fine until it suddenly feels awful.

Jitter is the variability of delay. A video call can tolerate some delay if it is steady, because buffers can compensate. It struggles with variable delay, because buffers either underflow (causing glitches) or grow too large (causing lag).

In the final mile, queuing delay and jitter are the villains more often than propagation delay. The distance from your house to your ISP's first router is small; the wait time inside a congested buffer can be huge.

An episode from the late 2000s gave this problem a name that spread beyond engineering circles: *bufferbloat.* In simple terms, bufferbloat happens when network devices—routers, modems, Wi-Fi access points—have large buffers that hold packets during congestion, and those buffers fill up. Large buffers were once considered an unambiguous good. If you can store more packets, you lose fewer. But the Internet's dominant transport protocol, TCP, interprets packet loss as a signal to slow down. Big buffers reduce loss, which can keep throughput high. The catch is that those packets are still waiting in line. The buffer becomes a parking lot, and the price of fewer crashes is more traffic jams.

The tragedy of bufferbloat is that it often punishes exactly the activities people care about: interactive ones. A household starts a big upload to cloud storage. The router buffers packets. The upload continues at a healthy rate, and the speed test looks fine. Meanwhile, the video call becomes delayed, because the small, time-sensitive packets are stuck behind a long queue of bulk traffic. The user blames the call provider, or the laptop, or the neighbor. The actual cause is the invisible queuing discipline inside a box on a shelf.

Bufferbloat is not a moral failure by hardware vendors. It emerged from reasonable design choices in a world where broadband speeds were rising rapidly and memory was cheap. It took years of measurements and advocacy to shift the industry toward actively managing queues rather than merely enlarging them. The underlying lesson is as old as cities: capacity without management can create its own form of congestion.

Nowhere is this clearer than in cable broadband, which in many countries has been delivered over infrastructure originally designed for one-way television distribution. The technology family that governs cable Internet is DOCSIS. It is a remarkable engineering achievement: delivering high-speed data over coaxial cable networks with amplifiers, taps, and shared segments. But its structure also illustrates the physics of shared access.

On many cable networks, the downstream (from provider to home) is a broadcast-like medium: many homes share a downstream channel. The upstream (from home to provider) is even more constrained, because multiple homes must take turns transmitting. That turn-taking is not a casual etiquette; it is scheduled. A cable modem cannot simply shout onto the line whenever it wants, or it would collide with its neighbors. Instead, it requests transmission opportunities and is granted them by the network's scheduler. Under load, the waiting time to get a grant can become part of the latency budget.

This is a different world from dedicated fiber to each home, where the physical medium is effectively private. Shared media can be efficient and economical, but it makes access latency and queuing behavior central. It also means your experience can be shaped by the behavior of others on the same segment. The neighbor who begins a large upload is not stealing your bits directly, but they are contributing to the shared contention and the scheduling pressure that determines when your packets get a turn.

Engineers have responded with mechanisms that aim to keep latency under control even when the link is busy. Active Queue Management (AQM) is a family of techniques in which the network deliberately signals congestion earlier, often by dropping or marking packets before buffers become huge. The counterintuitive idea is that a small amount of controlled pain prevents a larger, uncontrolled one. In the DOCSIS world, standards and operational practice have increasingly incorporated AQM requirements so that buffers in cable modems do not silently grow without bound. More recent efforts under the banner of *Low Latency DOCSIS* go further, attempting to isolate latency-sensitive traffic into a queue that is protected from the bulk-traffic queue, while also addressing the scheduling delays inherent in upstream request-grant systems. The goal is to make "busy" feel less like "broken."

The final mile is also where radio enters the story, even for people who think they are on "wired Internet." Wi-Fi is a local access network inside the home, and it is often the actual bottleneck. Wi-Fi is a shared medium with its own contention mechanisms. It competes with neighboring networks, with microwaves, with building materials, and with the simple fact that airtime is finite. A household can pay for a fast broadband plan and still experience poor responsiveness because the Wi-Fi access point is overloaded or poorly placed. From the user's perspective, the Internet is slow. From the network's perspective, the bottleneck is three meters away.

This creates a subtle social consequence: when performance depends on home equipment, responsibility becomes ambiguous. The ISP may claim the line is fine to the modem. The user may observe real problems on their laptop. Both can be correct. The hidden Internet, in this case, includes the hidden physics of a living room: where the router is placed, whether it is behind a television, whether it is competing with a neighbor's access point through a thin apartment wall.

Measurement helps turn vague frustration into diagnosis. One powerful approach is to look at round-trip time (RTT) to the first few hops outside the home. If the RTT to the first hop—often the ISP's gateway—swells dramatically under load, the problem is likely in the access link, the modem, or the home router. If the first hop stays stable but later hops become delayed, the congestion may be deeper in the ISP or beyond. Distributed measurement platforms that place probes at the edge, in ordinary networks, can make these patterns visible across geographies. When thousands of small vantage points report, the final mile becomes a measurable phenomenon rather than a personal anecdote.

A simple household experiment illustrates the point. Begin a large upload, such as sending a video file to cloud backup. At the same time, continuously ping a nearby, stable target (or run a real-time latency test). On a well-managed connection, the ping times rise modestly. On a bufferbloated connection, they can rise by hundreds or even thousands of milliseconds. Nothing has changed about the speed of light. The distance to the target is the same. The delay is pure waiting, created by how devices choose to buffer and schedule packets.

This is why the final mile is sometimes described as a problem of *responsiveness* rather than *speed*. Users do not live inside throughput graphs. They live inside feedback loops: click and response, speak and reply, press and see. A connection that is theoretically fast but practically laggy feels worse than a slower connection that stays responsive under load.

The economics of the final mile follow the physics. Because upgrades are expensive and localized, access networks are often built as layered compromises: a mix of old and new plant, a patchwork of neighborhoods upgraded at different times, equipment generations coexisting uneasily. The system can deliver impressive average performance while still producing pockets of chronic contention at peak times. Those pockets are not always profitable to fix quickly, especially if the market does not reward latency improvements as visibly as it rewards advertised download speeds.

Advertising has trained consumers to think in megabits per second. Physics keeps reminding engineers that milliseconds matter. In online gaming and video calling, a stable 20 ms is often more valuable than an unstable 200 Mbps. Yet it is harder to sell *less jitter* than *faster downloads*. The final mile therefore becomes a place where the incentives of markets, regulators, and engineering do not always align.

There is also a political geography to the final mile. Dense areas are cheaper to wire per customer; rural areas are expensive. Apartment buildings can be upgraded efficiently if building owners cooperate; single-family homes can require more labor. Rights-of-way and pole attachments can become bottlenecks in their own right, turning construction logistics into an invisible limiter of national connectivity. The Internet's egalitarian myth runs into the very unequal costs of digging.

Still, the story is not one of inevitable stagnation. Fiber-to-the-home deployments show that the final mile can be transformed when investment and policy align. Wireless technologies, including fixed wireless and newer cellular systems, can offer alternatives where trenching is prohibitive, though they bring their own spectrum and contention constraints. Improvements in queue management and scheduling can make existing infrastructure feel dramatically better without changing raw bandwidth. Even within a single household, a better-placed access point, a wired backhaul, or a router with sensible queue management can turn a brittle connection into a stable one.

What makes the final mile distinctive is that it is where the Internet touches bodies and buildings. Fingers tapping screens. Voices riding packets. Landlords signing access agreements. Technicians climbing poles in winter. When someone says the Internet is "down," they usually do not mean an undersea cable has snapped; they mean something has failed or congested in the part of the system that is closest, most physical, and most entangled with daily life.

In the end, the final mile is less like the grand arteries of global fiber and more like the capillaries of a nervous system: numerous, fragile, and essential. The core can be engineered for abundance, but the edge must be engineered for fairness under contention, for low delay under load, and for the unglamorous reality that a civilization-scale network still depends on what happens at the curb, at the basement demarcation point, and on the shelf where a small blinking box decides, packet by packet, who gets to go first.

Chapter 4

The Accidental Organism

We often imagine civilization's machinery as a blueprint drawn by a master architect, yet the reality is far messier and more organic. From the synchronization of power grids to the routing of internet traffic, our infrastructure exhibits behaviors-growth, healing, and panic-that owe more to biology than to engineering. This chapter peels back the concrete and silicon to reveal the adaptive, living system beneath.

4.1 The Symphony of Local Rules

The first impulse, when confronted with something as sprawling as the Internet, is to imagine a control room. A wall of screens. A master map. A few calm people steering the whole thing with joysticks and switches. That picture is comforting because it matches older machines: a train has a driver, a factory has a foreman, a telephone exchange once had operators plugging cables into boards. Surely the hidden internet—the nervous system of civilization—must have someone at the helm.

But the most important fact about modern infrastructure is that there is no such room.

Order appears anyway. Messages arrive. Videos play. Bank transfers clear. Navigation apps juggle road closures and still find you a route. This is not because the system is simple; it is because it is full of small decisions, made close to where the action is, obeying rules modest enough to fit on a page. The global pattern is the consequence, not the command.

A useful word for this is *emergence*. It does not mean "magic." It means that when many parts follow local rules, a large-scale behavior can appear that no part contains as a plan. A single ant does not understand "the colony's strategy," but the colony can build, forage, defend, and even regulate temperature. A single driver does not intend to create a traffic jam, yet jams bloom on highways like weather systems. A single router on the Internet does not know the "best" path across the planet, yet the network carries a pulse of data from any device to almost any other, most of the time.

The contrast becomes clearer if you sketch two ways to organize a system: the rigid hierarchy, and the fluid swarm.

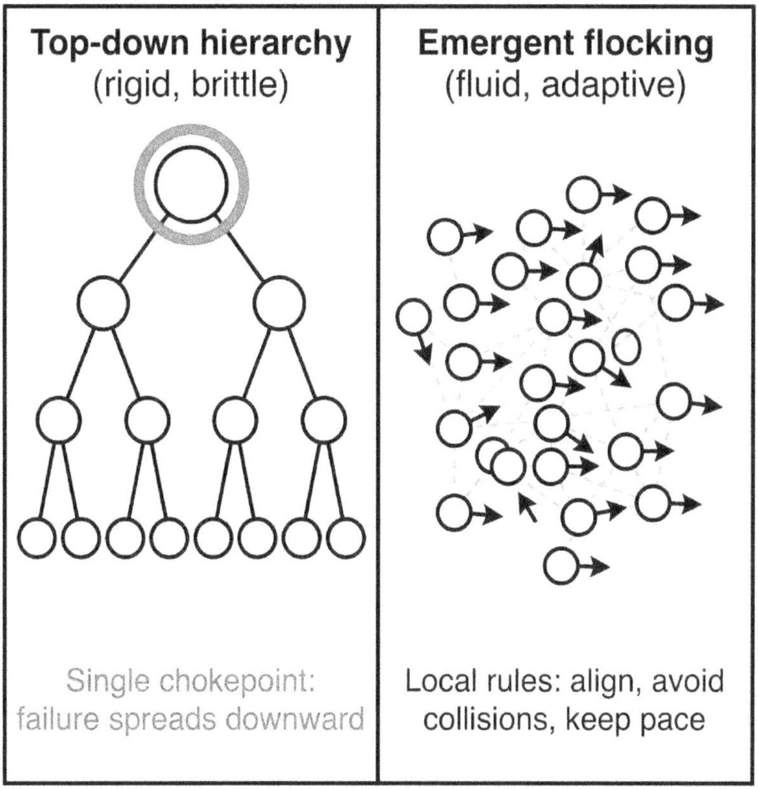

The left-hand picture is how we often talk about "control": one command becomes many instructions, flowing down. It is tidy, but it hides a weakness. The top is a chokepoint. If the root is confused, compromised, or simply overwhelmed, everything beneath inherits the failure. Hierarchies can be excellent when tasks are predictable and environments are stable, but they have a particular kind of fragility: the cost of being wrong is paid everywhere at once.

The right-hand picture is less tidy: a group of agents, each with a sense of its neighbors. No one issues a grand order. Yet the group moves coherently. It can bend around an obstacle. It can absorb the loss of a few agents. It can split and rejoin. The price of this adaptability is that the behavior is harder to "read" from any single part. The whole is not stored anywhere; it is enacted.

The Internet is far closer to the right-hand picture. Its coherence comes from protocols: agreements about how to behave locally. A protocol is not a law of nature and not a central plan. It is closer to etiquette, except that etiquette can be broken without consequence, while protocols are enforced by physics, software, and the silent refusal of machines to cooperate with nonsense.

Consider the humblest of these courtesies: the acknowledgement.

When a device sends data across the Internet, it is chopped into packets—small envelopes of bits. Those packets pass through routers, which are specialized machines that forward packets toward their destinations. The network makes no promise that any particular packet will arrive. A cable can be cut. A queue can overflow. A radio signal can fade. Instead of demanding certainty from the network, the endpoints adopt a local ritual: "I sent this; did you get it?" That answer is the acknowledgement, often abbreviated as *ACK*. If the sender does not receive an ACK, it tries again.

On paper, this seems almost childish. Surely civilization should not rely on a global nervous system that amounts to "tell me you heard me." And yet the simplicity is the point. Because the rule is local—because it can be carried out by the two endpoints without anyone's permission—it scales. Billions of devices can participate without a central registry of "who is talking." The global effect, reliable communication over unreliable paths, emerges from countless small acts of asking and answering.

The same logic holds for congestion, the Internet's version of traffic.

Roads are a familiar place to watch emergence at work, because we can see it with our eyes. You may have experienced the puzzling phenomenon of a "phantom traffic jam": the highway slows to a crawl, then, ten minutes later, it clears with no crash, no lane closure, no obvious cause. The jam is real, but it is not an object like a fallen tree. It is a wave—a collective pattern caused by local reactions. One driver taps the brakes. The driver behind brakes a little more. The next one brakes harder still. The ripple grows as it travels backward, and by the time it reaches you it feels like an external event: *traffic happened to me*.

Data networks have an equivalent. When too many packets reach a router at once, they form a queue. A small queue is useful: it smooths bursts. But if packets arrive faster than they can be forwarded, the queue grows until it overflows. Packets are dropped. If senders respond to those losses by sending even more aggressively, the situation becomes self-defeating: the network spends its effort handling packets that will be discarded, like a post office drowning under mail it cannot deliver.

A historical anecdote captures how close the Internet once came to this failure mode. In the mid-1980s, parts of the network experienced what engineers later called "congestion collapse," where throughput—the rate of useful delivery—fell dramatically even as everyone tried harder. The rescue came not from a central authority shutting people off, but from a change in a local rule: how each sender adjusts its rate in response to perceived congestion. Van Jacobson's work at Berkeley (1988) popularized a set of mechanisms—slow start, congestion avoidance, fast retransmit—that taught endpoints to treat loss and delay as signals. The idea is almost moral in tone: when the network seems strained, back off. When it seems healthy, cautiously increase. Each sender acts selfishly (it wants to deliver its own data), but it follows a protocol that makes selfishness compatible with the common good.

This is not a story about technological virtue. It is a story about incentives made mechanical. If you obey the rule, your traffic tends to get through. If you ignore it, you trigger losses that harm you first. That alignment is what allows a global pattern to settle: the Internet becomes usable not because everyone is benevolent, but because the local protocol makes restraint a winning strategy.

Even routing—the question of *where* packets go—works this way. There is no single map of the Internet that all routers consult. Instead, the network is a patchwork of independently operated domains called *Autonomous Systems*: Internet service providers, cloud companies, universities, enterprises, content delivery networks. Between these domains, a protocol called the Border Gateway Protocol (BGP) shares reachability information: "to get to these addresses, you can send traffic to me." BGP is sometimes described as the Internet's postal service, but it is more like diplomacy. Each Autonomous System has its own policies: which neighbors it trusts, which routes it prefers, which traffic it will carry for money, which it refuses on principle or regulation. Each makes local decisions. The world-wide routing state is the collective consequence.

That has an unsettling implication. If no one has the whole map, no one can guarantee that the outcome is globally optimal. The Internet does not compute *the* best possible routes. It computes *a* set of routes that are mutually compatible with local policies and current conditions. And because the conditions change—links fail, maintenance happens, providers renegotiate agreements—the computed world shifts. The network converges, wobbles, and reconverges, like a crowd finding paths through a doorway.

This is where the metaphor of a symphony earns its keep. In an orchestra, no musician hears the entire audience's experience directly. Each hears their neighbors and watches a conductor. The Internet has no conductor, but it does have a score: protocols define timing, phrasing, and permissible moves. The "music" is the emergent behavior: stability, reachability, acceptable performance. When the score is well designed, local competence produces a coherent whole. When the score is flawed, the same competence can generate chaos—precisely because everyone is reacting quickly and faithfully to local cues.

A concrete example of how delicate this can be is the infamous case of "too much trust" in routing announcements. In 2008, Pakistan's attempt to block access to YouTube accidentally leaked beyond its borders, and a portion of global Internet traffic for YouTube was briefly misdirected. The details are technical, but the moral is simple: the system was not defeated by a mastermind; it was tripped by an ordinary local action amplified by ordinary local obedience. Routers heard an announcement that sounded valid and did what they always do: they believed it and forwarded accordingly. The global pattern—worldwide misrouting—was an emergent consequence of local rules meeting an unexpected input.

It is tempting to treat such incidents as proof that decentralized systems are inherently unsafe. But centralized systems have their own catastrophes, and often more spectacular ones. A single flawed database can freeze an airline. A single mistaken software update can brick millions of devices. The point is not that local rules are always superior. The point is that, at the scale of civilization, local rules are often the only rules that can be executed quickly enough, widely enough, and cheaply enough to keep the system alive.

The same dance appears in power grids, which are not the Internet but rhyme with it. The alternating current grid must maintain a common frequency—roughly 60 Hz in North America, 50 Hz in much of the rest of the world. That frequency is not "set" by a single dial. It is an emergent property of many generators and loads, each pushing and pulling on the grid. Engineers sometimes describe this as synchronization: many oscillators locking together. When you switch on a kettle, you do not ask permission from a central dispatcher in real time. You simply draw power. The grid reacts. Generators, controllers, and protective devices respond locally—governors adjust mechanical input, inverters tweak output, relays trip under dangerous conditions. If those responses are tuned well, the frequency stays steady enough that you never think about it. The coherence is a continual achievement.

This way of thinking changes what "intelligence" means in infrastructure. Intelligence is not concentrated in a headquarters; it is spread across mechanisms that measure, decide, and act close to where information is freshest. A router does not need to understand geopolitics to drop packets when its queue is full. A TCP sender does not need to know who else is online to slow down when acknowledgements arrive late. A generator governor does not need to know tomorrow's weather to respond to a frequency dip right now.

Once you start seeing it, this distributed intelligence is everywhere. A supermarket supply chain has reorder points: if stock drops below a threshold, reorder. That local rule can stabilize shelves—until it synchronizes too many actors and produces a shortage wave. A hospital has triage protocols: treat the most urgent first. That local rule can save lives—until a surge overwhelms capacity and forces a grim redefinition of "urgent." A social media feed has engagement rules: show what users react to. That local rule can make feeds feel personalized—until it amplifies outrage and misinformation. The same basic ingredient, local decision-making, can bake either resilience or fragility depending on the feedback it creates.

The Internet's local rules are unusually literal: they are encoded. That makes them powerful, because they are consistent. It also makes them perilous, because a mistake is not a misunderstanding between humans; it is an executable misunderstanding between machines. When a protocol says "back off when you see loss," it creates a kind of peace treaty among senders. When a protocol says "prefer shorter routes," it creates a bias that can be exploited by a false announcement. When a protocol says "cache this answer for a while," it speeds up naming at the cost of making the world's directory slightly out of date.

The great achievement of the hidden internet is not that it eliminates conflict among actors; it is that it channels conflict into forms that do not immediately tear the system apart. Providers compete. Users binge. Attackers probe. Hardware fails. Undersea cables are cut by anchors. Yet most days, the network behaves like a coherent organism. Not because it has a brain, but because it has reflexes.

There is a final, counterintuitive lesson in all this. When coherence is emergent, you cannot fully understand the whole by interrogating a single part, no matter how diligently. You can examine a router's configuration, a congestion-control algorithm, an interconnection contract, a fiber map—and still be surprised by what happens when millions of such pieces interact at speed. That surprise is not always a sign of incompetence. It is often the price of building systems large enough to serve everyone, fast enough to feel instantaneous, and flexible enough to survive constant change.

A civilization that depends on such systems becomes fluent in a new kind of trust: not trust in a ruler, but trust in rules. The everyday miracle is not that the rules are perfect; it is that they are simple enough to be shared, local enough to be executed, and compatible enough to compose into something that, from the outside, looks like intention.

4.2 Echoes and Amplifiers

A small child learns the idea of feedback by accident. You push a swing; it comes back. You push again at the right moment; the arc grows. Push at the wrong moment; the motion dies. No one needs to explain the mathematics. The body learns that timing matters, and that a system can either calm itself down or run away from you depending on how you respond to its own response.

Infrastructure behaves like that, except the "push" might be a factory starting its motors, a million people refreshing a website after breaking news, or a trading algorithm reacting to a price tick. The response might be automatic and invisible: a relay tripping, a router changing its route, a server spinning up new instances, a human operator dispatching generation. The loop closes when that response changes the environment that triggered it. That circularity is what engineers and biologists alike call *feedback*.

Feedback can be stabilizing. A thermostat is the household parable: when the room gets cold, the heater turns on; when it gets warm, it turns off. The thermostat does not "know" comfort in any human sense. It enforces a rule that resists drift. Engineers call this *negative feedback*, not because it is bad, but because it subtracts from deviations. It is a constant, quiet refusal to let the system wander.

Feedback can also be destabilizing. A microphone too close to a speaker produces that piercing squeal: a bit of sound enters the mic, gets amplified, emerges from the speaker, re-enters the mic louder, and so on until the system saturates. That is *positive feedback*: it adds to deviations. It turns a small difference into a big one.

Modern infrastructure is thick with both kinds. The hidden internet, the power grid, roads, markets, and media are not static constructions but dynamic organisms of sensors, controllers, and reflexes. Their metabolism is made of loops.

To see how a stabilizing loop feels from the inside, consider electricity. Most people meet the grid as a wall socket: a promise of power that seems as steady as gravity. But the grid is a balancing act performed in real time. Supply must match demand almost instantaneously. If millions of kettles click on, generators must push more mechanical energy into spinning rotors; otherwise frequency dips. If a large generator trips offline, the system must shed load or recruit reserves; otherwise frequency falls far enough to trigger protective shutdowns.

The first line of defense is local and automatic. Generator governors respond to changes in frequency: if frequency begins to drop, they open valves or adjust fuel to increase mechanical input. That response is not a centralized command; it is a reflex built into each machine. Many such reflexes, acting together, produce a global effect: the grid tends to return toward its nominal frequency. This is negative feedback as homeostasis, an organism maintaining internal conditions despite external perturbations.

A different stabilizing loop lives in the Internet's basic transport protocols. When congestion rises, packets are dropped or delayed; the sender interprets that as a sign to slow down. When congestion eases, the sender cautiously speeds up. Each flow acts locally, and the global effect-when the parameters are right-is a network that shares capacity without collapsing into useless retransmissions. It resembles a crowd that learns, without speaking, to stop pushing when the doorway clogs.

Yet the same ingredients can flip signs. A loop that stabilizes under ordinary conditions can become an amplifier under stress, or when the response is delayed, mismeasured, or misaligned with what is actually happening.

Delays are particularly dangerous because they turn correction into overcorrection. Imagine steering a car with a laggy wheel: you turn, nothing happens, you turn more, then the car suddenly swerves. The easiest way to create a runaway system is to combine strong reactions with late information.

Economists have long watched this phenomenon in supply chains. A small change in consumer demand can create large oscillations upstream: the retailer orders more, the wholesaler orders even more to be safe, the manufacturer ramps up, and by the time the goods arrive the demand has normalized-leaving warehouses stuffed. That wobble is known as the *bullwhip effect*. It is not caused by malice or incompetence; it is the natural product of local actors responding to local signals with delays and uncertainty. Infrastructure, too, can whip itself.

The Internet has its own bullwhips, some of them built from good intentions. One of the most telling examples involves BGP, the routing protocol that allows independently run networks to tell one another which destinations they can reach. BGP is often described in terms of topology-who connects to whom-but its day-to-day reality is temporal. Routes appear, disappear, change preference, and propagate across the planet. When something breaks, the network does not instantly "know" the best new configuration. It gropes toward one by exchanging updates. Engineers call the period of adjustment *convergence*.

In the late 1990s and early 2000s, operators faced a real problem: some routes were "flapping," repeatedly switching between available and unavailable because of an unstable link or misconfiguration. Every flap triggers updates that consume CPU, memory, and bandwidth on routers. Enough flapping can degrade the network for everyone, like a crowd constantly changing lanes.

A proposed cure sounded like common sense: if a route changes too often, penalize it. Suppress it for a while so that the network calms down. This family of techniques became known as *route flap damping*. The loop is straightforward. You observe instability; you respond by reducing that route's influence; you hope to reduce instability.

In practice, this negative feedback sometimes acted like a positive one. The key problem was interaction with BGP's own behavior during failures. When a route is withdrawn, routers do not instantly leap to the new best path. They may explore alternatives, briefly advertising and retracting intermediate options. A damping system can interpret those transient explorations as repeated "flaps" and decide to punish a route that is, in fact, part of recovery. The punishment then delays convergence, prolonging instability and creating more churn-exactly what damping was meant to reduce. The network's attempt to heal is mistaken for sickness.

There is something almost tragic in this. A system designed to quiet noise can, under particular conditions, amplify it by silencing the very signals needed for coordination. The lesson is not that damping is foolish, but that feedback loops rarely exist alone. In real infrastructure, loops are coupled: one controller is embedded inside another, and their interactions matter as much as their individual logic.

Power grids offer a more dramatic version of the same story. Cascading blackouts can begin with a small disturbance-one line trips, one tree touches a conductor, one operator action goes wrong-and then unfold through a chain of protective responses that were each reasonable in isolation. A relay trips a line to prevent overheating. That shifts power to other lines, increasing their load. Those lines trip to protect themselves. Generation becomes separated from load. Frequency deviates. More protection triggers. The system fragments.

Protective devices are, in essence, local feedback controllers. They watch a measurement-current, voltage, frequency-and act when it crosses a threshold. Their purpose is to prevent local damage. But local protection can conflict with global survival. In biology, an immune response that is too strong can harm the host. In a grid, protection that is too eager can create the very collapse it seeks to avoid.

Statistics adds another twist. Large blackouts are not simply scaled-up versions of small ones. Work by Ian Dobson and collaborators (2007), looking at blackout series statistically, helped popularize the idea that grid behavior can show heavy-tailed risk: a small number of events account for a large fraction of total impact. That shape is familiar from floods and earthquakes. It also appears in networks and markets. When risk is heavy-tailed, the question "What's the average outage?" is almost beside the point, because the average is dominated by rare extremes.

One reason heavy tails appear is that complex systems can drift toward criticality. In everyday language, "critical" means "dangerous." In the language of dynamical systems, a critical point is a boundary between regimes: on one side, disturbances die out; on the other, they propagate. Near the boundary, the system can look stable for long periods, and then surprise you with a large cascade triggered by an otherwise ordinary shock. The unsettling part is that the system can *self-organize* toward this boundary. If operators invest heavily to suppress small failures, they may inadvertently allow stress to accumulate—higher loads, tighter margins—until the system is poised for a larger one. This is not a moral accusation against operators; it is a reminder that feedback exists at the level of institutions too. Policies respond to outcomes, and those responses change the conditions that produce the next outcomes.

The Internet shows analogous behavior in digital epidemics and flash crowds. Consider a news event that drives millions of people to the same page. Each person is not trying to create an outage. But the collective effect can overload servers, saturate upstream links, and trigger automatic defenses. A content delivery network might reroute traffic; a cloud autoscaler might spin up new machines; a DDoS protection service might begin challenging clients. These responses are often stabilizing, and when they work they are almost invisible.

When they misfire, they can create amplification loops. If a site begins returning errors, users refresh. Refreshing increases load, producing more errors, prompting more refreshing. Social media can pour fuel on that loop by broadcasting the site's unavailability, which sends yet more people to check whether it is back. The loop is not located in a single device. It is distributed across human habits and machine policies, stitched together by attention.

That entanglement of humans and machines matters because it can flip the sign of feedback. A purely technical controller might have been designed with negative feedback in mind. Add a human layer, and the effective loop can become positive. A bank is stable when depositors trust it. A rumor of trouble can prompt withdrawals. Withdrawals make the trouble real. That is a run, and it is a feedback loop driven by belief. The hidden internet carries belief at high speed.

Even apparently simple mechanisms like caching-storing answers so they can be reused-contain feedback. The Domain Name System (DNS), which turns names into numerical addresses, relies heavily on caching so that the same lookups are not repeated endlessly. When caches are warm, the system is fast and efficient; the load is spread out. When caches expire simultaneously, a spike of queries can hit the authoritative servers, causing delays or failures that prevent caches from refilling, causing more repeated queries. Engineers call this kind of synchronized demand a "thundering herd." It is an amplifier created not by any one bad actor but by many actors doing the same sensible thing at the same time.

A subtler amplifier lives in the way networks respond to loss. Congestion control algorithms interpret loss as a sign of congestion and slow down, which is stabilizing. But some applications are impatient: they open multiple parallel connections, retry aggressively, or switch to different protocols to maintain responsiveness. From the user's perspective, the application is "trying harder." From the network's perspective, the application is multiplying itself. Under certain conditions, such behavior turns congestion into more congestion, especially when errors are interpreted as "send again, quickly."

This is why feedback loops feel like metabolism rather than machinery. In a clock, gears do not change their ratios because the clock is running late. In a living organism, tissues do change their behavior because the organism is hungry, stressed, or injured. Infrastructure is increasingly the latter. It senses and reacts. It remembers, in the blunt way that software and policy remember: a timer set, a penalty accumulated, a threshold adjusted.

The difference between echo and amplifier is often a matter of calibration: how strong the response is, how quickly it happens, how accurately it measures what it claims to measure, and how it interacts with other loops. A home thermostat that is too sensitive will chatter on and off. A power-grid controller that reacts too aggressively can excite oscillations. A routing stabilization mechanism can suppress recovery. A misinformation "correction" can unintentionally spread the misinformation further by repeating it. The loop is not just a circle; it is a circle with gain.

The most practical way to think about all this is to stop asking, "What is the component supposed to do?" and start asking, "What does the component do to the system when everyone else is also doing their part?" A brake pedal is meant to slow a car. A million brake pedals can create a traffic wave. A single route penalty can quiet a flapping link. Millions of local penalties can delay a global reconvergence. A single protective relay can save a transformer. Thousands of protective relays can fragment a grid.

None of this means infrastructure is doomed to collapse. Negative feedback is real and powerful. The world does not reboot every morning. Most of the time, the loops do exactly what they are meant to do: they dampen. They stabilize. They turn shocks into ripples. Their success is the quietest thing about them.

But the moments when the loops go the other way-when an error echoes into a roar-reveal the true nature of the hidden internet. It is not a static utility delivered from on high. It is a living fabric of responses, woven from protocols, devices, institutions, and human impulses, always listening to itself and always, in one form or another, replying.

4.3 Quarantining the Collapse

A child learns the appeal of modularity without ever using the word. Blocks click together into a tower, and the tower falls. The satisfying part is not that it fell; it is that the pieces remain pieces. You do not have to melt the plastic back into a single lump to rebuild. You pick up what survived, swap a few pieces, and try again.

Civilization's hidden systems are expected to behave the same way. When a fault occurs, the hope is not always to prevent it—many faults are inevitable—but to keep the fault from turning into a total-body failure. The technical term is *containment*. The biological metaphor is immunity. In both cases, survival depends on boundaries: membranes that let useful flows pass while slowing or stopping dangerous ones.

That sounds simple, even obvious, until you remember what these systems are made of: electricity, money, data, and trust. All of them are prone to contagion. A surge on one line can destabilize an entire interconnection. A mistaken route announcement can redirect traffic worldwide. A vulnerability in a common software library can become a near-universal infection. When everything is connected, why does anything ever stop spreading?

The answer is that connectivity is not a single state. There are many kinds of connection, and the details matter. A cell is full of chemical reactions, but it is bounded by a wall. A ship is made of watertight compartments so that a breach does not sink it immediately. A city has fire doors and firebreaks. The Internet has *Autonomous Systems*-administrative domains with their own rules, contracts, and control planes-stitched together by protocols that assume disagreement and delay. These boundaries are not perfect, but they are the difference between an accident and an extinction.

Begin with a piece of maritime engineering that is almost too on-the-nose: the bulkhead. When the Titanic struck an iceberg in 1912, its designers had included watertight compartments intended to limit flooding. The tragedy is often told as a morality play about hubris, but it is also a lesson in modularity's specifics. The ship did have compartments; the flood did not stay in them. Water spilled over the tops as the bow sank, turning a series of local breaches into a progressive, system-wide failure. A boundary that looks solid in a blueprint can be porous in the dynamics of an actual disaster.

That story generalizes. Modularity is not the presence of walls; it is the ability of walls to remain effective under stress. A boundary must be paired with a rule about when to close it, how to route around it, and what costs will be paid to preserve the rest of the organism.

Power grids provide a stark example because the "fluid" being contained—electrical power—cannot simply be stored in pipes. The grid's coherence depends on frequency. When generation suddenly falls short of load, frequency drops. If it drops too far, generators and industrial motors can trip, worsening the imbalance. The system can enter a death spiral.

One of the least romantic but most consequential firebreaks in the grid is *Underfrequency Load Shedding* (UFLS). The phrase sounds bureaucratic because it is. UFLS is a codified last resort: if frequency falls below certain thresholds, predefined blocks of load are automatically disconnected. This is not failure; it is controlled injury. The grid sacrifices a limb to keep the heart beating.

From the user's perspective, UFLS looks like an outage with no warning. From the grid's perspective, it is a tourniquet. It buys time for generators to recover and for operators to rebalance the system. Like many immune responses, it is blunt. It does not ask which customers are most deserving; it follows preset schemes because speed matters. The moral discomfort—who loses power first?—is part of why the mechanism is planned in advance rather than improvised under duress.

There is a deeper, less comfortable point: UFLS makes the system survivable precisely because it admits that the system will sometimes be pushed beyond stable operation. Feedback loops can stabilize within a regime; modularity is what keeps the organism alive when it leaves that regime. In other words, there are two kinds of safety. One is keeping the car on the road. The other is having crumple zones when it doesn't.

The Internet's equivalent of UFLS is not a single switch but a collection of boundaries and validations that prevent certain classes of mistakes from becoming universal truths. Internet routing is especially instructive because it is both intensely technical and intensely social. Routing decisions are influenced by economics, politics, contracts, and grudges, yet they are executed by routers that obey protocols with no capacity for diplomacy.

The basic unit of modularity in Internet routing is the Autonomous System (AS). An AS might be an Internet service provider, a cloud operator, a university network, or a large enterprise. Each AS controls its internal routing and then exchanges reachability information with neighboring ASes using BGP. This architecture is often described as "decentralized," but it is more precise to call it *federated*. Each domain is sovereign within its borders. The Internet is what happens at the seams.

Those seams can act like bulkheads. An AS can refuse certain routes, filter announcements, or set policies that keep trouble from crossing its boundaries. That is not merely an option; it is how the system acknowledges reality. Not every network trusts every other network. They cannot. Even honest mistakes—misconfigurations, fat-fingered prefixes, accidental leaks—are common enough that blind trust would be suicidal.

A modern example of engineered containment is *Route Origin Validation*, typically built on the Resource Public Key Infrastructure (RPKI) and Route Origin Authorizations (ROAs). The core idea is simple enough to explain at a dinner party: if a network claims it is the legitimate origin for a block of IP addresses, other networks can check that claim against signed records. If the claim is invalid, the route can be rejected or de-preferred. This does not solve every routing security problem, but it can prevent a specific class of accidents and attacks: an unauthorized network originating a prefix and attracting traffic that does not belong to it.

There is a subtlety here that matters for the "organism" metaphor. Validation depends on distributed data that may not be perfectly consistent everywhere at once. Different operators may fetch validation information from different caches; caches may lag; policies differ. This is not a flaw so much as a recognition that global consistency is expensive and slow. The design accepts partial, local knowledge and tries to make it safer than naive trust. Modularity is not only a boundary; it is an admission that knowledge is local, and that safety must work anyway.

Containment also appears in less formal ways. Content delivery networks (CDNs) are a kind of modularity for performance and resilience. They replicate content across many locations so that the failure of one data center does not eliminate access. But they also create compartments: if one region is overloaded, traffic can be diverted; if one cache is poisoned, it can be flushed without collapsing the entire system. The modular unit is no longer an AS but a point of presence, a cluster, a shard of cached state.

The most vivid "quarantine" stories, however, come from malware and worms, because they behave like pathogens in the folk sense. They spread. They exploit contact. They reveal the price of homogeneity.

In 1988, the Morris worm spread across what was then a much smaller Internet, exploiting vulnerabilities and misconfigurations in common services. Its author reportedly did not intend to cause the scale of disruption that followed, but intention is irrelevant to a contagion. The worm replicated, machines slowed or crashed, and administrators scrambled. One reason this event remains famous is that it occurred when the network community was small enough to react as a community; another is that it exposed how quickly a shared weakness could become a shared emergency.

Since then, the scale has grown, and so has the reliance on shared software components. A vulnerability in a widely used library can be the biological equivalent of a receptor present on every cell: a single exploit can suddenly have billions of potential hosts. In such a world, containment depends on being able to segment networks, limit privileges, and roll out patches quickly. The language of "zero trust" and "least privilege," while fashionable, describes ancient immune principles: do not assume friendliness; minimize what any one infection can reach.

Network scientists have long pointed out that not all networks behave the same under contagion. In a fairly uniform network—where each node has roughly the same number of connections—random immunization can be effective because the "average" node matters. But many real-world networks, including parts of the Internet, are heterogeneous: a small number of nodes (hubs) have vastly more connections than most. Romualdo Pastor-Satorras and Alessandro Vespignani's work (2001) helped popularize the idea that such networks can sustain spreading dynamics very differently from homogeneous ones, and that interventions focused on hubs can be disproportionately powerful.

The social meaning of this is awkward. It implies that a small number of organizations—major providers, large platforms, key infrastructure operators—can strongly influence whether a contagion becomes a pandemic. That influence can be a public good when exercised responsibly and a systemic risk when it fails. Modularity helps here too: if the network can be arranged so that no single hub is a universal hub, or so that hubs are protected by stronger boundaries, the organism is less prone to catastrophic infection. But there is a tradeoff. Hubs exist for reasons: efficiency, economies of scale, low latency, convenience. The same structures that make the system fast and cheap can make it susceptible.

Containment is also political. In a pandemic, quarantine measures are not purely medical; they are decisions about freedom, work, and fairness. Infrastructure quarantine is no different. When a network operator filters routes aggressively, it may protect its users but harm reachability for others. When a grid sheds load, it protects the interconnection but imposes costs on those disconnected. When a platform rate-limits users to stop abuse, it also limits legitimate use. The "immune system" is always also a governance system, deciding whose pain is acceptable to spare whom.

This is where earlier ideas about local rules and feedback loops become practical rather than philosophical. A boundary that never closes is not a boundary. A boundary that closes too eagerly is suffocation. The immune system that attacks every foreign particle is autoimmune disease. The firewall that blocks too much is a self-inflicted outage. The art is to create compartments that are permeable under normal conditions and selectively impermeable under threat.

A final image helps anchor the point: forest fires. Firebreaks—roads, rivers, cleared strips—are modularity in landscape form. They do not prevent ignition. They prevent every ignition from becoming a megafire. In recent years, fire scientists and land managers have had to confront the way suppression itself can alter the system: put out every small fire, and fuel accumulates; when the inevitable big fire arrives, it burns hotter. The same uneasy dynamic appears in engineered systems. Suppressing small failures can, under some conditions, move the system into a more brittle regime. Containment must be paired with honesty about stress, maintenance of margins, and periodic release of pressure in controlled ways.

The hidden internet survives because it is full of deliberate weak points—places designed to give way first. A circuit breaker that trips. A rate limiter that drops requests. A routing filter that refuses an announcement. A load-shedding relay that darkens a neighborhood to keep a continent synchronized. These are not signs of fragility. They are signs of a system that expects to be hurt and has decided, in advance, where it can afford the injury.

In the best moments, modularity is almost compassionate. It limits the blast radius. It makes disasters smaller than they might have been.

It preserves enough functioning tissue that recovery is possible without rebuilding the organism from scratch. The deeper achievement is not that collapse never happens, but that collapse is so often local—confined to a compartment, a subnet, a feeder line, a region—while the larger body keeps its pulse.

Chapter 5

Choreography of Atoms

The device you are reading on arrived via a massive, silent dance of global coordination, where software commands steel and electricity. We peel back the user interface to reveal the heavy machinery of civilization, demonstrating how the logic of the internet has terraformed the physical logistics of our world.

5.1 The Physical Packet

A glass of wine on a table carries an invisible tour of infrastructure: a vineyard's tractors and pumps, a bottling line, a carton, a pallet, a truck, a port, a ship, another port, a warehouse, another truck, and finally a shop shelf. When it arrives intact, it feels as natural as gravity. Yet for most of industrial history, moving things across long distances was not a smooth tour at all. It was a series of awkward handoffs, each one a chance for delay, damage, theft, and paperwork errors. The hidden internet of civilization is not only made of fiber and radio waves. It is also made of steel corners and standardized holes, of cranes designed to grab the same geometry everywhere on Earth. Containerization did something startlingly similar to what packet-switching did for data: it turned a continuous, messy physical world into discrete, addressable units that can be routed, queued, and recombined.

Before the container, ports were theaters of improvisation. Cargo arrived as break-bulk: sacks of coffee, barrels of molasses, crates of machine parts, coils of rope, bundles of timber, each with its own handling quirks. Dockworkers built temporary sculptures of freight in ship holds, fitting irregular shapes together like three-dimensional puzzles. A ship's time at berth was dominated not by sailing but by loading and unloading, and that time was expensive. The labor was heavy, the record-keeping often manual, and the losses could be substantial. Even when everyone involved was honest and competent, break-bulk made it easy for a shipment to become an argument: a missing crate could be a dockside mistake, a warehouse miscount, a ship's hold tally gone wrong, or a theft that left no clear evidence.

The key to understanding the container is to treat it not as a box but as a protocol. A protocol is a shared rulebook that lets different parties cooperate without having to know each other's internal details. On the digital internet, TCP/IP works because it defines what a packet looks like and how devices should treat it. A router does not need to know whether a packet contains a love letter, a bank transfer, or a software update; it needs to read the header, decide where to send it, and move it along quickly. Containerization pulled off a related trick in the world of atoms. A crane operator does not need to know whether a container holds sneakers, electronics, or wine; the equipment needs to engage the same corner castings, lift within a known weight limit, and put the box onto a ship, a railcar, or a truck chassis built for the same dimensions.

This sounds obvious now because the container has become part of the background, as unremarkable as the shape of a light switch. But the simplicity is the achievement. Standardization does not merely make things tidy; it changes what can be automated, what can be insured, what can be planned, and what can be trusted at scale. In the same way that a standardized electrical plug makes it possible to buy appliances without negotiating with your wall, the standardized container makes it possible to build a global logistics system where handoffs between companies and countries can be routine rather than bespoke.

The historical hinge is often told through the story of Malcolm McLean, an American trucking entrepreneur who, in the mid-1950s, pushed a radical idea: instead of unloading a truck's cargo piece by piece, lift the entire cargo box off the truck and onto a ship. The first voyages of containerized cargo were not instantly a world system; they were experiments that demanded new ships, new port equipment, new labor agreements, and new regulatory arrangements. The brilliance was not simply inventing a box. It was insisting that the box be treated as the unit of transport across modes, a single "thing" that could move from road to sea to rail without being opened. That insistence forced a cascade of redesigns. Ports that had been built around teams of men carrying and stacking suddenly needed gantry cranes, container yards, and scheduling systems. Ships that had been designed around irregular stowage had to become floating grids. Rail networks had to adapt to double-stacking and specialized well cars. Trucking needed chassis and depots designed for quick interchange.

If this were only a story of efficiency, it would already be big. But the deeper change is that a standardized unit makes a network behave differently. Imagine a city without standardized mail envelopes. Every letter would be a different size and shape; postal sorting machines would be useless; delivery routes would be slow because each item would require special care. The container is the envelope for physical trade. It does not eliminate complexity inside the envelope, but it moves that complexity to the edges: to factories that pack, to warehouses that pick, to retailers that unpack. The transport network in the middle becomes more uniform and therefore more mechanical, more predictable, and more scalable.

This is where the "physical packet" metaphor earns its keep. A packet in networking is a bundle: payload plus the information needed to move it. The payload is the data; the header carries addresses and other control information. A shipping container similarly separates cargo from transport. The cargo can vary wildly, but the container's external form stays constant. And while a container does not carry its full "header" painted on the outside, it does have a kind of physical identity designed to be read by systems: markings, codes, seals, and associated documents. Modern logistics lives in a constant conversation between the steel box and the data about the box. The identity is what lets the world treat the container as a unit that can be tracked, allocated space, and transferred between custodians.

International standards formalized this identity. ISO container standards define classes, dimensions, and ratings so that equipment built in one country can handle boxes built in another. Equally important are identification standards such as ISO 6346, which specify a coding system and markings intended for documentation and control, including automatic data processing. In other words, a container is designed not only for cranes and ships but for databases and messaging systems. The box is a node in an information network as much as it is a piece of metal.

A concrete example makes the abstraction tangible. Consider a container of specialty coffee moving from Colombia to a roaster in the United States. At the origin, the coffee may be bagged and placed on pallets, then loaded into a container at a facility that can secure it, document it, and seal it. The container door is closed with a seal that functions like a tamper-evident checksum: it does not prevent opening, but it makes unauthorized opening detectable. The container is then drayed by truck to a port terminal. The terminal does not handle individual bags of coffee. It handles containers, each one with a known footprint and weight, each one addressable in the terminal operating system. The yard is laid out as a grid, like memory in a computer, and containers are stacked in bays. When the ship arrives, a crane lifts containers in a planned sequence, often optimized to reduce rehandles and to keep the ship balanced. The ship is not loaded with "coffee"; it is loaded with container slots, each filled by a container whose contents are, to the ship, irrelevant. On arrival, the process reverses: lift, stack, retrieve, move by truck or rail, deliver to a warehouse, open at the edge of the network where the coffee is once again treated as coffee rather than as a standardized block of volume.

The striking part is what did *not* have to happen in the middle. The coffee did not have to be touched by dockworkers in two ports. It did not have to be tallied as individual sacks in a ship's hold. It did not have to be repacked to switch modes. It did not have to be negotiated as a bespoke object at every handoff. The container imposed an abstraction layer: transport deals in containers; commerce deals in goods. As in computing, an abstraction layer can be life-changing because it allows different parts of the system to evolve independently. A coffee producer can change packaging or pallet configuration without requiring a redesign of ships. A port can upgrade its cranes without needing every exporter to change how they bag beans. The agreement lives at the interface: the container's dimensions, strength, corner fittings, and identification.

Once a standard unit exists, something like "packet-switching for atoms" becomes imaginable. Packet-switching in data networks relies on breaking communication into small packets that can travel different routes, be buffered at intermediate nodes, and be reassembled at the destination. The physical world cannot compete on speed, and goods cannot be reassembled without labor. Yet the network logic rhymes. Containers allow relay-like movement through hubs: a container can be offloaded at a transshipment port, stacked temporarily (buffered), then loaded onto another vessel (switched) headed elsewhere. Rail terminals and inland depots play similar roles. The routing decisions are not made by a single omniscient authority; they emerge from schedules, contracts, capacity, weather, congestion, and disruptions. When a port is jammed, containers wait in a yard like packets in a queue. When a vessel is late, downstream connections are missed and containers are rerouted, like packets taking alternate paths when a link fails. The resemblance is not perfect, but the key idea holds: standardization makes it possible to build a general-purpose transport network in which intermediate nodes can handle units without understanding their contents.

The metaphor becomes even more revealing when one looks at "headers" and "payloads." A TCP/IP packet header includes source and destination addresses, sequencing information, and error detection. A container's "header" is distributed across markings and accompanying data: the container identification code, the shipper, the consignee, the port of loading, the port of discharge, the hazardous goods classification if relevant, the weight, the seal number, the booking reference, the bill of lading, and customs filings. Much of that header is not printed on the container, but the container is the anchor that ties the physical and informational together. Scan the ID and you can pull the rest from a database. That is why the container's code is engineered to be legible to machines, and why logistics companies invest heavily in the systems that keep that mapping accurate.

This also explains a subtle point: containerization did not eliminate paperwork; it changed its role. In break-bulk, paperwork was often a fragile shadow of the physical pile. In containerized shipping, the paperwork becomes a kind of control plane, coordinating a physical data plane of boxes moving through ports and across oceans. When the control plane is correct, the flow can look effortless. When the control plane is wrong, you can have a steel box that is perfectly safe, perfectly sealed, and perfectly in the wrong place.

There is a human story behind this shift. In the era of break-bulk, dock labor was a highly skilled and often tightly organized workforce. Knowing how to stow cargo so it would not crush itself, shift in heavy seas, or become unreachable was a craft. Containerization moved much of that craft into engineering and planning: the ship is designed with cell guides, the port with cranes and yard layouts, the loading plan with software. For dockworkers, the change could mean fewer jobs, different skills, and labor conflict. For shippers, it meant fewer losses and faster turnaround. For consumers, it meant that the world's variety could appear in local stores at prices that would once have seemed implausible. The container's quiet steel corners are a social technology as much as a mechanical one, reshaping work and bargaining power alongside costs.

A historical anecdote that captures the magnitude is the transformation of port landscapes. Old urban waterfronts once devoted to piers, sheds, and finger docks became obsolete as container terminals demanded vast flat expanses for stacking and truck access. In many cities, container ports migrated away from downtown to locations with room for yards and highway and rail connections. The waterfront changed from a dense, labor-rich interface between city and sea into a high-throughput machine on the edge of town. That migration is visible in real estate maps and skyline views, but it is also visible in the economic geography of trade: once a container can be transferred seamlessly between ship and rail, inland locations become more competitive, and supply chains can be designed with different balances of labor, land costs, and access.

Standardization has another consequence that is easy to miss: it encourages modular thinking. A container can hold pallets; pallets can hold cases; cases can hold individual items, each layer with its own identifiers. Organizations like GS1 standardized product identifiers such as the Global Trade Item Number (GTIN), along with barcodes and RFID schemes that allow automated capture and sharing of information. This nested stack of standards means that a modern supply chain can "see" at multiple resolutions. At one moment the system cares about a container; at another, a pallet; at another, a single retail item. The container is the coarse-grained packet, efficient for long-distance bulk movement. Finer-grained identifiers allow sorting and fulfillment closer to the customer. The result is a multi-layer network, like the layered architecture of the digital internet, where different layers optimize different tasks.

A minimalist schematic helps clarify the structural similarity between a data packet and a container without pretending they are identical.

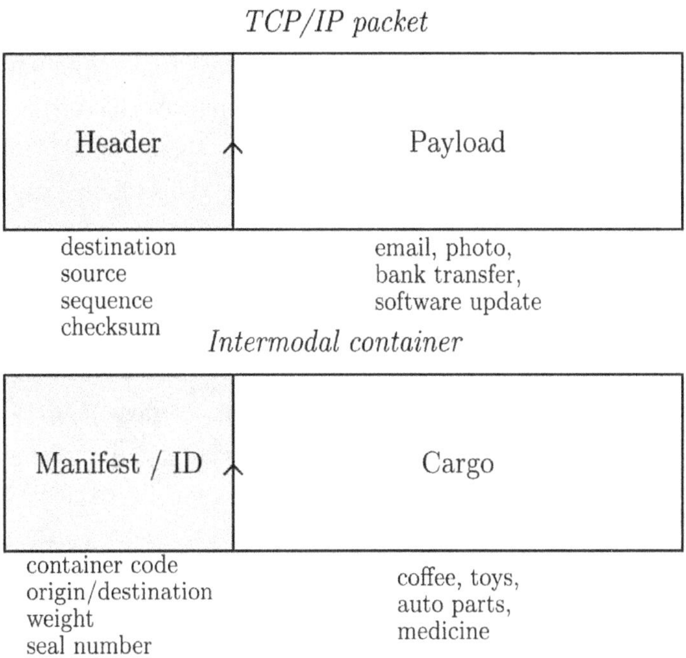

Same structural idea: a standardized container plus standardized information enables routing without inspecting contents at every handoff.

The point of the diagram is not to claim that shipping containers behave exactly like IP packets. A data packet is tiny, fast, and disposable; a container is large, slow, and valuable. Packets can be copied perfectly; containers cannot. Packets can be rerouted in milliseconds; containers may wait days. Yet the shared structural idea matters: when the transport system can operate on standardized units, it becomes possible to build high-capacity hubs and predictable interfaces, and to let many independent actors coordinate through agreed formats rather than through constant negotiation.

That coordination is, in practice, what makes the container feel like a natural fact rather than a diplomatic miracle. A container might be manufactured in one country, owned by a leasing company in another, loaded by an exporter, carried by an ocean carrier, handled by terminal operators in two ports, inspected by customs authorities, placed on a rail service run by a different company, and delivered by a trucking firm. None of these organizations share a single internal system or a single set of incentives. They cooperate because the unit is standardized and because the information about the unit can be exchanged in standardized ways. Interoperability is not a warm feeling; it is a discipline enforced by geometry, codes, and contracts.

The container also reshaped risk. When cargo is sealed inside a steel box, it is protected from weather and casual pilferage, but it is also opaque. A sealed container is a privacy technology for commerce, which is good for efficiency and sometimes good for safety, yet it also creates new challenges for inspection and security. Customs agencies and port authorities have had to develop screening methods that can cope with opacity at scale: selective inspections, risk scoring, scanning technologies, seal verification, and data-driven targeting. The better the control-plane information, the more selectively a system can inspect without clogging the network. When that information is absent or unreliable, inspection becomes either lax (with obvious risks) or heavy-handed (with obvious delays).

A final detail brings the "nervous system" metaphor into focus: the container is not only standardized in size. It is standardized as a *symbol* in many systems at once. Financial systems care about it because it represents value in motion. Insurance systems care about it because it concentrates risk. Scheduling systems care about it because it consumes scarce space on ships, trains, and in yards. National security systems care about it because it crosses borders. Environmental systems care about it because it enables trade volumes that shape emissions and land use. When a unit becomes standardized, it becomes countable; when it becomes countable, it becomes governable; when it becomes governable, it becomes a lever that institutions fight over.

The romance of global trade is often told through maps and ship silhouettes. The container's quieter romance lies in the fact that it made the world's movement legible to machines. A steel box with a code, a seal, and a known geometry can be lifted by a crane that has never "met" it before. A yard plan can treat it as a coordinate. A ship can treat it as a slot. A rail terminal can treat it as a block in a train. A warehouse can treat it as an arrival event with a predicted time. When that works, the box seems to glide through infrastructure as if civilization had always been built this way: goods in standardized packets, moving through hubs, switched and buffered and routed, the choreography of atoms written in steel.

5.2 When Code Moves Steel

A container terminal looks, from a distance, like a simple landscape: water, concrete, cranes, stacks of steel boxes. Up close it behaves more like a living instrument. Vehicles weave with the patience of insects. Cranes move with a deliberation that is almost graceful, each lift ending in a click of geometry as a container's corner castings settle into a spreader. Lights change. Gates open and close. Somewhere in a building that could be mistaken for an office, a cluster of screens shows a diagram of the yard divided into rectangles, each rectangle representing a place where a real, heavy object may or may not be sitting.

The physical packet made global trade modular, but modularity only matters if the modules can be coordinated. The coordination happens in a layer that is neither purely digital nor purely mechanical: industrial control. In heavy industry, software does not merely *describe* the world. It *pushes* it. A line of code becomes a voltage change; a voltage change becomes a valve position; a valve position becomes the flow of water or gas or grain or molten metal. This is the moment where the hidden internet touches the earth hard enough to leave marks.

The umbrella term is Industrial Control Systems (ICS): the computers, networks, and devices used to monitor and control industrial processes. A closely related term is SCADA, for Supervisory Control And Data Acquisition. The words are revealing. *Supervisory* means there is a level above individual machines where humans and software oversee the process. *Control* is not metaphorical: it is the ability to start, stop, speed up, slow down, open, close. *Data acquisition* means sensors constantly telling the supervisory layer what is happening right now. The system is a loop: measure, decide, act, measure again.

It helps to separate two ideas that are easily confused. A business system tells you that a container is booked, that a truck is arriving at 14:10, that an order must ship today. An industrial control system can make the gate barrier rise, can energize the motor that moves a conveyor, can command a crane to limit its speed when the wind is high. Both are "software", but one lives in the world of promises and records, the other in the world of torque and inertia. The first can ruin a balance sheet; the second can break a machine or a person.

At the heart of this translation sit devices with deceptively humble names. A sensor is a device that turns a physical quantity into an electrical signal: temperature, pressure, vibration, position, flow rate, level in a tank, presence of an object, even chemical composition. An actuator is the opposite: a device that turns an electrical signal into physical action. An actuator might be a valve that opens a pipe, a relay that switches power, a motor drive that changes speed, a brake that engages, a switch that changes a railway track. Between sensors and actuators sits a controller that decides what to do. In modern industry that controller is often a PLC, a Programmable Logic Controller: a rugged industrial computer designed to run continuously, tolerate electrical noise, and interact with the messy realities of industrial wiring.

If TCP/IP made it possible to move information in packets without caring what the information meant, a PLC makes it possible to move *intent* into machines without caring how the machine is built internally. The PLC reads a set of input signals, applies logic, and writes outputs. That logic can be as simple as "if the emergency stop is pressed, cut power," or as subtle as "if the tank level is rising too quickly, close the inlet valve gradually while starting the pump to avoid pressure shock." The most revealing feature of PLC logic is that it is rarely written like the software in a phone app. It is often expressed in ladder logic, a diagrammatic form inspired by electrical relay circuits. The metaphor is not accidental. PLCs replaced panels of physical relays. The code is a new kind of wiring.

The supervisory layer above PLCs is where SCADA comes into view. A SCADA system collects measurements, displays them to operators, logs them for analysis, and can send commands down to the controllers. In a water treatment plant, the operator might see the level of a reservoir, the status of pumps, the chlorine dosing rate, alarms that signal abnormal conditions. In a power grid control room, operators see frequencies and voltages, line loadings, generator outputs, and protective system status. In a container terminal, the equivalent supervisory functions are often handled by a mix of systems: terminal operating systems for planning and record-keeping, and industrial control for the cranes, gates, and safety interlocks. The result can feel like a single unified machine, but it is often a federation of subsystems, stitched together by networks and standards and the improvisations of engineers.

A feedback loop diagram makes this concrete. The essential structure is always the same: digital controller, actuator, physical process, sensor, back to the controller. The loop runs continuously, sometimes many times per second. When it is well-designed, the loop keeps the process in a safe and efficient range despite disturbances. When it is poorly designed, the loop can oscillate, overshoot, or drift into dangerous territory. When it is attacked or misconfigured, the loop can become a weapon against itself.

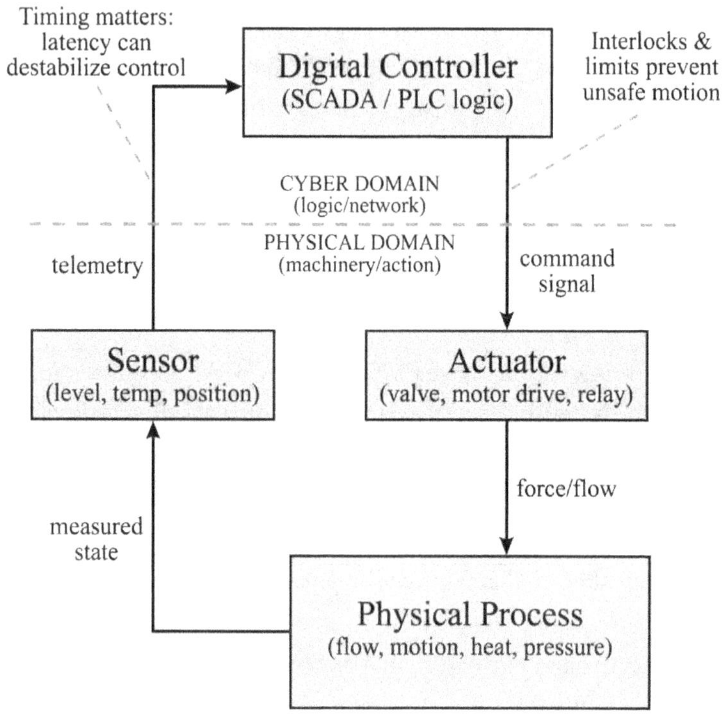

The loop explains why industrial systems are both powerful and peculiar. In office software, a glitch is irritating. In control software, a glitch can become kinetic. Consider something as everyday as a conveyor belt in a distribution center. The belt must start and stop at the right times so boxes do not collide, so workers are not endangered, so scanners can read labels accurately. The controller watches photoelectric sensors that detect the presence of a box, then decides whether to run the motor, slow it, or divert the box down another path. Multiply that by thousands of belts, diverters, and scanners, and the warehouse becomes a mechanical computer: physical objects are the "data", and the conveyor network is the "bus" that carries them. The code is the scheduler.

The stakes rise further in systems that manage energy and fluids. A refinery is essentially a city of pipes. A power plant is a controlled conversion of heat into rotation into electricity. A railway signaling system is an agreement between metal wheels and distance: trains can only safely share a line if switches and signals coordinate their motion. Each domain has its own hardware, but the control logic shares a family resemblance: closed-loop feedback, alarms, safety thresholds, redundancy.

This is where the hidden internet becomes counterintuitive. One might imagine industrial control as a separate world, isolated and proprietary, like a sealed cockpit. Historically that was closer to true. Older control systems were physically separated, using dedicated wiring and specialized protocols. Over time, economics and convenience pulled industrial systems toward general-purpose networking. Ethernet cables are cheap. IP networks are flexible. Remote monitoring saves travel time. Vendors offer dashboards that can be viewed from laptops. Data from machines can be aggregated for predictive maintenance. The logic of connectivity is relentless: if information can be captured, it will be captured; if it can be transmitted, it will be transmitted.

Yet connectivity changes the threat model. The moment a controller can be reached over a network, it becomes part of a wider world of software vulnerabilities, misconfigurations, and malicious actors. Heavy industry, meanwhile, has constraints that make security harder than in typical IT environments. Many controllers are designed to run continuously for years. Applying patches can require planned outages. Some equipment uses legacy protocols that were never designed with authentication or encryption in mind, because the designers assumed physical isolation. In an office network, if a laptop needs rebooting, it is an inconvenience. In a port or a power plant, a reboot might halt throughput or trip a process into a safe shutdown that takes hours to restart.

A concrete example of the collision between cyber incidents and physical operations came in 2017 with the NotPetya malware outbreak, widely reported as having caused massive disruption to Maersk, one of the world's largest shipping companies. The details that made headlines were financial, but the operational picture was visceral: bookings disrupted, systems down, terminals struggling to process cargo. In a world of physical packets, a terminal can sometimes keep moving containers for a short time by using manual workarounds: radios, paper notes, ad hoc instructions. But the yard is not a simple pile of boxes. It is a carefully indexed memory. If you lose track of which container is where, the terminal begins to behave like a computer that has forgotten where it stored its files. The cranes can still lift, but each lift risks turning order into entropy. Throughput collapses not because steel is weak, but because the information needed to move steel efficiently has become unreliable.

The terminal is a useful place to see how many layers of control co-exist. A crane itself is a sophisticated machine with its own internal controllers. It must manage sway, limit speed, prevent overloading, and enforce safe zones. Those functions are often handled by embedded systems close to the machinery. Above that, there may be a yard control system that tells the crane which container to pick and where to place it. Above that, there is planning software that decides which containers should be retrieved first to meet ship schedules and trucking appointments. Above that, there are commercial systems handling bookings, bills of lading, customs documentation, and customer updates. A disruption at any layer can propagate, but the propagation is not symmetrical. If the crane's internal controller fails, nothing moves. If the planning layer fails, movement becomes less efficient and may become unsafe. If the commercial layer fails, the machinery may still move, but the movement risks losing its meaning: boxes may arrive without the digital "headers" that tell the rest of the world what those boxes are and where they should go.

This layered view also clarifies an essential concept in control engineering: safety is often handled by a separate, more conservative system than optimization. A well-designed industrial site includes hardwired emergency stops, safety relays, mechanical brakes, pressure relief valves, and protective devices that do not depend on a complex network being healthy. These are the physical equivalents of fail-safes. They are designed to bring a process to a safe state even if higher-level software becomes confused. The presence of such safety layers is not a sign that engineers distrust software; it is a recognition that complex systems fail in complex ways, and that physics does not negotiate.

The translation from code to steel is not only about keeping things from going wrong. It is also about squeezing extraordinary performance out of machines. A modern automated terminal can choreograph vehicles and cranes to achieve impressive throughput, minimizing idle time. A steel mill can keep temperatures and flows within tight tolerances, improving quality and reducing waste. A power grid can balance supply and demand in real time, keeping frequency stable even as wind gusts and cloud cover change generation. The human operator is not removed from these systems so much as repositioned. Operators become supervisors of automation, intervening when anomalies appear, responding to alarms, and making judgment calls when the system confronts a scenario it was not designed to handle.

That repositioning creates its own psychological hazards. When things run smoothly, humans can become complacent, because the system rarely asks for help. When the system fails, it may fail fast, presenting a flood of alarms and requiring rapid diagnosis under stress. Aviation has studied this pattern extensively; heavy industry lives with it daily. The hidden internet's nervous system is efficient, but efficiency often means fewer opportunities to practice manual skills.

A small, domestic analogy helps. Think of a thermostat controlling a home furnace. It senses temperature, compares it to a setpoint, turns heat on or off, and checks again. Now imagine a thermostat that also knows the price of electricity minute by minute, the weather forecast, the occupancy of rooms, and the health of the furnace. It could save money and energy. It could also, if hacked or misconfigured, make the house unlivable or even unsafe. Scale the house up to a city's water system or a port that feeds supermarkets, and the stakes become collective.

The most distinctive feature of industrial control, compared with the digital internet's familiar services, is the tyranny of time. In many control loops, milliseconds matter. Latency and jitter (variation in delay) can degrade stability. A command that arrives too late can be worse than no command at all. Engineers talk about deterministic behavior: knowing, within tight bounds, when a message will arrive and when an action will occur. General-purpose networks can be fast, but they are not always deterministic, especially when congested. This is why industrial networks often use specialized configurations, priority schemes, or protocols designed for real-time performance. It is also why uncontrolled connectivity can be dangerous: a flood of traffic that would merely slow an office network can, in extreme cases, interfere with time-sensitive control communications.

None of this means that industrial systems are fragile toys. They are built to survive harsh environments and human error. The problem is that their strength invites reliance, and reliance invites tight coupling. When a port depends on digital coordination to achieve competitive throughput, it cannot simply "go manual" for long. When a water plant depends on remote telemetry to optimize pumping schedules, losing that telemetry forces conservative operation that can waste energy and reduce resilience. The hidden internet does not only connect people to information; it connects sensors to decisions and decisions to actuators, and the loop runs whether or not anyone is watching.

The phrase "code moves steel" can sound like a slogan until you notice it everywhere. A rail signal changes, and hundreds of tons of train obey. A protective relay detects a fault, and a circuit breaker opens with a violent snap, isolating a segment of the grid. A crane control system enforces an invisible boundary, and a swinging container stops short of a ship's bridge. These are not metaphors. They are the everyday miracles of control engineering: logic translated into force, information translated into motion, the nervous system of civilization flexing its muscles with each correctly timed command.

When everything works, the physical world seems to have acquired a new property: responsiveness. Machines become less like dumb tools and more like coordinated limbs. A terminal can "know" that a truck is arriving and stage the right container. A factory can "know" that a motor is vibrating abnormally and schedule maintenance before a breakdown. A grid can "know" that demand is rising and bring generation online. The knowledge is not human understanding in any rich sense. It is a structured, operational awareness produced by sensors, protocols, and control loops. It is enough to move steel safely, quickly, and in the right direction, which is a kind of intelligence that matters even when no one is thinking about it at all.

5.3 The Brittle Clockwork

The container made freight behave like a standardized unit; industrial control made yards and factories respond to signals with mechanical obedience. Together they created a kind of physical internet that is astonishingly fast for matter. The cost of that speed is that the system learns to live with less slack. A little redundancy feels, from the inside, like waste: extra inventory sitting idle, unused warehouse space, spare trucks, additional suppliers who charge more, extra time baked into schedules that makes everything seem slower on paper. Competitive pressure rewards the opposite posture: run hot, run lean, keep everything moving.

Just-in-Time manufacturing, usually shortened to JIT, is the most famous expression of this posture. The slogan sounds like common sense: receive components only when needed, avoid piles of inventory, reduce storage costs, reveal problems quickly. Behind it sits a deeper redefinition. Inventory, in an older industrial mindset, was a kind of insurance policy. You bought protection against uncertainty by holding extra parts. JIT treats inventory as a buffer that hides inefficiency. Remove the buffer, the reasoning goes, and the system is forced to become more precise.

That precision can be beautiful. It can also be brittle. A tightly coupled system is one where parts depend on each other's timing with little room for delay. In a loosely coupled system, a local failure stays local because there is slack: alternative routes, extra stock, spare capacity, time cushions. In a tightly coupled system, a small interruption can cascade, not because anyone is foolish, but because the design trades resilience for throughput.

A useful way to see this is to borrow language from networks. Every transport network has queues: ships waiting at anchor, trucks in a gate line, containers stacked in a yard, work-in-progress piled between factory steps. Queues are not always bad. They can be stabilizing buffers, absorbing variation. But queues cost money and space. The temptation is to shrink them until the system feels perfectly efficient. In digital networks, shrinking buffers too far can lead to congestion collapse or jitter that ruins performance. In supply chains, shrinking buffers too far can turn ordinary variability into crisis.

The variability is not exotic. Demand changes with weather, holidays, social trends, and the simple unpredictability of human choice. A factory machine breaks. A truck driver calls in sick. A storm closes a port. A customs inspection takes longer than expected. A component arrives with a defect and must be reworked. Each event is manageable in isolation. The trouble begins when many links have been optimized under the assumption that the rest of the chain will behave close to schedule.

The signature symptom of tight coupling is the mismatch between how a system looks in a spreadsheet and how it behaves in real life. On paper, a chain with minimal inventory and high utilization looks superb: low carrying costs, high throughput, short lead times. In reality, high utilization leaves little room to absorb shocks. If a machine is scheduled at 95% capacity, a small disruption creates a backlog that cannot be cleared without overtime or delays. If a port yard is near full, a late ship can force chaotic reshuffling. If a supplier's production is tuned to a narrow window, a missed delivery can halt an assembly line in hours.

The world learned this lesson repeatedly in the early 2020s, when COVID-19 did not simply reduce supply or demand in a uniform way. It whiplashed them. Lockdowns suppressed some consumption while boosting other categories; factories closed and reopened unevenly; labor availability changed; transportation capacity shifted. What looked, from a distance, like a single "pandemic shock" was in fact a series of timing mismatches that moved through networks. The delays were not merely delays. They were phase shifts: a factory that restarted without enough parts, a retailer that ordered aggressively to compensate, a port that became congested, a ship schedule that slipped, a warehouse that ran out of space, a truck shortage that amplified everything. The system behaved less like a smooth pipeline and more like a clockwork mechanism in which one gear slipping throws off many others.

One of the most vivid anecdotes of chokepoint fragility is the grounding of the *Ever Given* in the Suez Canal in March 2021. A single ship, stuck sideways in a narrow passage, halted a major artery of global trade for days. The drama was easy to photograph: a colossal container ship immobilized by sand and physics, with tiny excavators working at its bow like toys. The deeper story unfolded in spreadsheets and scheduling rooms. Ships queued at both ends. Some rerouted around the Cape of Good Hope, adding time and fuel. Container equipment ended up in the wrong places: too many empty boxes in one region, too few in another. Cargo that was supposed to arrive in a steady rhythm arrived in a lump, stressing ports and inland transport. The blockage did not merely pause trade; it rearranged it in time, creating a wave of congestion that traveled long after the ship floated free.

A chokepoint, in network terms, is a high-betweenness link: many routes depend on it. In physical logistics, chokepoints include canals, straits, major ports, rail junctions, key bridges, and even specific factories that dominate the supply of a specialized component. Efficiency tends to concentrate flows through chokepoints because concentration creates economies of scale. Bigger ships prefer fewer ports. Large ports attract more services. Specialized factories outperform general ones. These are rational choices locally. Globally they create single points of failure that can be hard to notice until something breaks.

Buffers once softened chokepoint shocks. If a canal closes but warehouses are full, the immediate effect is muted. If warehouses are lean, the closure becomes visible as shortages within days. In consumer life, this translates into a strange and unsettling phenomenon: the absence of things that everyone assumed were always available. It feels like a failure of modernity. Often it is not a failure of production capacity so much as a failure of timing.

The timing problem becomes more counterintuitive when the system's information signals are distorted. The classic name for this distortion is the bullwhip effect. Imagine a retailer who sells 100 units of a product per week. Demand fluctuates a little: 98, 102, 101. The retailer tries to keep shelves full, so it orders from a distributor. If the retailer becomes nervous about stockouts, it may order more than sales suggest, adding "safety stock" on top. The distributor sees the retailer's orders, not the final customer's purchases, and makes its own decisions. If it also fears shortages, it orders even more from the manufacturer. Upstream, the manufacturer observes large swings in orders and ramps production up and down, even though consumer demand was relatively stable. A small wiggle at the end becomes a large wave at the beginning, like a whip crack traveling from handle to tip but in reverse: the tip is the factory.

The bullwhip is not simply a matter of irrational panic. It can arise from ordinary practices that make sense locally. Orders are often placed in batches to reduce shipping costs. Promotions and price changes cause customers to buy in bursts. Forecasting algorithms react to recent sales. When supply is constrained, buyers may inflate orders in the hope of receiving at least some allocation. Each behavior is defensible; together they amplify variance. When a chain is tightly coupled, amplified variance collides with minimal buffers, and the result is oscillation: overordering followed by overcorrection, inventory gluts followed by shortages.

These oscillations can be seen even in non-crisis times in industries like fashion or consumer electronics, where demand is inherently volatile. They are also visible in more sobering contexts. Hospitals experienced acute shortages of protective equipment early in the COVID-19 pandemic not only because global capacity was inadequate, but because supply chains were optimized for routine consumption patterns and had limited surge capacity. When demand spiked, the chain could not instantly stretch. Factories cannot conjure new production lines overnight; shipping capacity cannot expand without ships, containers, crews, and port labor; quality assurance and regulatory steps cannot be skipped safely. A system that looks "efficient" under normal conditions can be shockingly unresponsive under extreme conditions.

The word *clockwork* suggests a machine that is precise and reliable. The hidden fragility is that clockwork depends on a stable environment. A mechanical watch is accurate because its parts interact in predictable ways with little variation. Introduce dust, shock, or temperature extremes, and precision becomes vulnerability. Supply chains are a kind of socio-technical clockwork: machines, software, contracts, human labor, and geography all meshed. Their precision is purchased by assumptions about the environment. When the environment shifts abruptly, the same tight tolerances that made the system fast make it prone to seizing.

Another kind of brittleness comes from standards and software dependencies, which feel abstract until they break. The container itself is standardized, but the global system around it is not static. Standards evolve: identification schemes, documentation formats, customs requirements, and safety rules are revised. Even seemingly small changes can ripple. If an identification standard changes, databases and scanners must be updated. If a customs message format is revised, trading partners must coordinate upgrades. In digital systems, protocols can be updated by deploying new software versions. In physical logistics, deployment is slower because it crosses organizations with different budgets, timelines, and constraints. The result is an ongoing global negotiation in which interoperability must be maintained while everything changes underneath.

Here the earlier idea that "the header matters as much as the payload" becomes a warning. A physical shipment can be perfectly intact and still be effectively lost if the identity and documentation layers fail. A container in the wrong yard position is a nuisance; a container that cannot be matched to the right documents can be a legal and operational headache. When a port is congested, the temptation is to move boxes wherever there is space. That short-term relief can create a long-term search problem, increasing rehandles and slowing everything further. The system begins to behave like a computer with a fragmented disk: data is present but costly to retrieve.

A concrete illustration can be found in the way modern retailers replenish shelves. Many operate with rapid turnover and minimal backroom storage. The store is not meant to be a warehouse; it is meant to be a display and sales point. Replenishment depends on frequent deliveries, sometimes daily. That works when transport and distribution centers run smoothly. When a distribution center is short-staffed or a storm disrupts trucking, shelves can empty quickly. The consumer's experience is immediate and emotional: the missing item becomes evidence that something bigger is wrong. The retailer, meanwhile, may respond by placing larger orders to compensate, feeding the bullwhip upstream. Scarcity becomes contagious.

The same dynamics play out in manufacturing at higher stakes. An automobile plant is an expensive machine in its own right, and stopping it is costly. If it runs JIT with minimal component inventory, a missing batch of semiconductors or wiring harnesses can idle thousands of workers and machines. The plant might keep running by building incomplete vehicles and parking them until parts arrive, but that creates a new kind of inventory: unfinished goods that consume space and complicate scheduling. The buffer returns, but in a distorted and expensive form.

Resilience, then, is not merely "having more inventory." It is having the right slack in the right places, and the ability to reconfigure when conditions change. Sometimes that means multiple suppliers in different regions, even if the cheapest supplier is tempting. Sometimes it means keeping strategic stock of critical components with long lead times. Sometimes it means designing products to be more substitutable: using common parts across models, qualifying alternative materials, avoiding single-sourced bespoke components. Sometimes it means building visibility so that disruptions are detected early. Visibility is the informational counterpart of slack: knowing where things are and what is likely to arrive allows earlier, less frantic adjustments.

Visibility is also where the hidden internet becomes ethically and politically charged. To reduce buffers safely, firms invest in surveillance of their own flows: scans at every handoff, real-time tracking of vehicles, predictive analytics on delays, automated replenishment driven by point-of-sale data. These systems can make supply chains more responsive, but they can also increase dependency on a small number of platforms and intermediaries. A world where inventory is held "in the network" rather than "in the warehouse" is a world where network failures have material consequences. It is also a world where bargaining power accrues to whoever controls the data about flows.

Chokepoints are not always geographical. They can be organizational. A single software provider can become a chokepoint if many terminals or carriers rely on the same system. A single identity service can become a chokepoint if it is necessary for gates to operate. A single customs portal can become a chokepoint if it is the only legal path for clearance. In such cases, the bottleneck is not concrete or waterway; it is an interface. When that interface fails, steel waits.

The most unsettling aspect of brittle clockwork is how quickly it makes ordinary life feel uncertain. Modern urban life assumes that a pharmacy will have antibiotics, that a supermarket will have fresh produce, that a hardware store will have basic materials, that a repair shop can get parts. These assumptions are not guaranteed by nature; they are guaranteed by a choreography that runs continuously in the background. When that choreography is disrupted, the shortages do not merely inconvenience consumers. They reveal how interdependent society has become, how thoroughly the physical internet has been optimized for normality.

There is, however, a quiet competence in how often the system recovers. After disruptions, ports add shifts, carriers revise schedules, manufacturers redesign products to use alternative parts, retailers ration and substitute, governments release reserves, and engineers reroute flows. The clockwork is brittle not because it is poorly built, but because it is built for speed. Its fragility is the shadow of its precision.

A striking way to feel this is to stand near a distribution center late at night, watching trailers back into loading bays in a steady cadence. Each trailer is a promise that tomorrow morning shelves will look normal. The building hums with conveyors and scanners, and the flow feels inevitable. It is anything but. The normality is manufactured moment by moment, with buffers shaved thin and time windows tightened, until the entire system runs like a perfectly tuned mechanism that can tolerate small errors but not large surprises. In a world that is increasingly full of large surprises, resilience begins to look less like waste and more like wisdom.

Chapter 6

Anatomy of a Crash

We often imagine infrastructure failure as a spectacular physical explosion, but the reality is more often a silent, synchronized collapse of logic. By tracing the path from a minor localized glitch to a catastrophic system reset, we reveal how the very mechanisms designed to protect our networks-interconnectivity and redundancy-can inadvertently provide the pathways for their destruction.

6.1 The Chokepoint Paradox

Every large technical system contains places where the messy variety of the world is briefly forced through something narrow. Those narrow places are not always obvious. Sometimes they are physical: a handful of fiber routes that run through the same duct, a substation that feeds a dense business district, a landing station where an undersea cable meets the shore. Sometimes they are logical: a small set of servers that translate names into numbers, or a routing policy that decides which path "counts" as the Internet today. In calm weather, chokepoints feel like clever engineering. Concentration makes things efficient: fewer boxes to manage, fewer contracts to negotiate, fewer moving parts to misbehave.

Then one day, concentration stops being efficiency and becomes leverage. A *Single Point of Failure* (often shortened to SPOF) is a component whose failure can stop an entire service. The phrase sounds like a scolding from an engineering checklist, but the deeper idea is unsettling: a SPOF is often created not by neglect, but by success. A system that works well tends to attract more dependence. Dependence thickens around whatever is reliable, cheap, fast, or centrally located. Over time the system grows a "center of gravity," and the center becomes the place where a small local problem can turn into a large public event.

A useful way to picture this is to imagine a city's road network. You can have many streets and still have one bridge that everyone uses because it is the only convenient way across the river. The map looks redundant: lots of lines, lots of options. Yet in practice the traffic converges. The same thing happens in the hidden Internet and in the electrical grid beneath it. The lines may look like a mesh, but incentives and physics can coax them into behaving like a hub-and-spoke.

A chokepoint can be a machine, a building, a cable, a protocol, a configuration file, or a contract. A surprising number of them are not owned by governments or by mythical "the Internet," but by ordinary companies optimizing for ordinary reasons. This is why chokepoints are so hard to reason about in advance. Nobody sits down and votes to make a particular DNS provider, transit network, or cloud region a linchpin. It simply emerges from the accumulation of rational choices.

The paradox begins with the most intuitive fix. If one thing is critical, add a backup. If one path carries most of the traffic, provision a second path. If one data center is a risk, replicate into another. Redundancy is the first word in every resilience playbook, and for good reason. Yet redundancy often changes the *shape* of failure rather than removing failure. It can trade a clean break for a confusing cascade. It can turn a simple system you understand into a coupled system you do not. It can even produce the special kind of fragility where the backups fail *because* they exist, or because they activate at the same time.

To see why, it helps to separate two questions that are easy to blur.

First: *Do you have an alternate component?* That is redundancy in the narrow sense. Second: *Can you safely switch to it under stress?* That is where the paradox lives. The switch itself has to be coordinated, and coordination is a hidden dependency.

The simplest example is human. A hospital may keep a backup generator, but if staff rarely test it, the backup is a guess. A telecom operator may have a disaster recovery plan, but if it lives in a binder and depends on the one person who knows the routing quirks, it is not really redundant. Technical redundancy that depends on rare human heroics is not redundancy; it is a delayed accident.

In networks, the coordination problem is often automated, which sounds comforting until you remember that automation is fast and literal. It will faithfully execute the wrong idea at scale.

Consider the naming system that makes the Internet usable for humans. People do not type IP addresses; they type names. The Domain Name System (DNS) is the distributed directory that turns *example.com* into a number that packets can route to. DNS is designed to be distributed and cached, and that distribution is real. Yet many organizations concentrate their DNS into a managed provider because it is cheaper, more reliable on average, and easier to operate. A managed DNS provider is not "the Internet," but if enough popular sites use the same provider, the provider becomes a chokepoint that can make unrelated services appear to vanish at once.

The famous illustration is the 2016 attack on Dyn, a company that provided managed DNS. A large distributed denial-of-service attack flooded Dyn's infrastructure, and because many high-traffic websites relied on Dyn for name resolution, users experienced something closer to a general Internet failure than a single-company outage. The websites themselves were often still running; the problem was that many users could not *find* them. The failure was not at the "content" layer but at the "direction signs" layer.

The episode is sometimes told as a morality play about insecure consumer devices, because the botnet involved many compromised cameras and routers. But the deeper lesson is architectural: a system can be globally distributed at the packet level and still be dependent on a small set of coordinating services. DNS is not a mere accessory. It is a steering wheel.

What does redundancy look like here? Many organizations responded by adding a second DNS provider. That is sensible, but notice how quickly the easy part ends. Running two DNS providers is not like having two light bulbs. You must keep their records consistent, handle differences in features, manage credentials and access controls for two separate platforms, and ensure that your automation does not "helpfully" update one while forgetting the other. You must also consider *how* clients will split or fail over between them. DNS caches can cling to old answers. Some resolvers behave differently from others. A small mistake in a record can persist long after you think you have fixed it, because caches are designed to remember.

A common fantasy is that redundancy is simply multiplication: two providers means half the risk. In reality it can become addition: two providers means two sets of knobs that can be mis-set, and a new category of failures where they disagree. The system gains a new failure mode: *inconsistency*. Inconsistency is not dramatic like a cable cut; it is worse. It produces partial outages, geography-specific weirdness, and symptoms that look like user error. The help desk cannot reproduce what the customer sees. Monitoring shows green in one region and red in another. Meanwhile, the organization is burning time not on repair, but on *diagnosis*.

The same pattern appears in the routing system that binds the Internet's many networks together. Between major networks, routing is coordinated by the Border Gateway Protocol (BGP). At a high level, BGP is how one network tells another, "I can deliver traffic to these IP prefixes; send it to me." It is decentralized and policy-driven: each network chooses which routes to accept, prefer, or export, often based on business relationships as much as on performance.

This flexibility is part of what made the Internet scale. It is also what makes the Internet sensitive to mistakes and to misplaced trust. A single incorrect announcement can propagate outward because downstream networks tend to re-advertise what they learn. A route that is *more specific* (a smaller block of addresses) often wins over a broader one, because it appears to be a more precise promise. That mechanism is useful when it is legitimate; it is dangerous when it is not.

A concrete example unfolded in February 2008, when Pakistan Telecom attempted to block YouTube domestically by announcing a more specific route to YouTube's addresses and directing it into a black hole. The plan may have been intended to affect only local users, but the announcement leaked beyond Pakistan's borders via upstream providers and spread. Much of the world began sending YouTube traffic into Pakistan, where it disappeared. YouTube was, for many people, simply gone for a time. A small local policy turned into a global misdirection.

That incident is often described as a hijack, and it was in effect, but it also illustrates chokepoints created by *control planes*. The control plane is the layer that tells the data plane what to do. Packets might be flowing along many physical paths, but a control-plane error can redirect them all toward a single bad destination. The chokepoint is not a cable; it is the trust relationship embedded in routing policy.

Redundancy here looks like multiple upstream providers, multiple peering links, perhaps even multiple data centers. Yet if the same wrong route is accepted by all of them, redundancy does not help. Worse, redundancy can accelerate spread because there are more conduits for propagation. A route leak can fan out through several well-connected networks more quickly than through one.

This is the first uncomfortable point: chokepoints are not always "one box." They can be a widely shared *decision*. When many actors use the same rule of thumb, the rule becomes a bottleneck.

That brings us to the second uncomfortable point: redundancy can create tight coupling, and tight coupling is what turns manageable failures into system-wide ones.

Coupling is a plain word for a profound fact. If two components must agree to function, they are coupled. If they fail at the same time for the same reason, they are coupled. If they recover in a way that stresses each other, they are coupled. Coupling can be created by shared software, shared vendors, shared geography, shared power feeds, shared time sources, shared certificate authorities, shared update mechanisms, shared staff, or shared assumptions.

Engineers often distinguish between *independent* redundancy and *correlated* redundancy. Independent redundancy is what you want: two different things that are unlikely to fail together. Correlated redundancy is the trap: two things that look separate but share a hidden dependency.

A classic correlated dependency is geography. Two data centers in the same city may look like redundancy on a spreadsheet, but they may share the same floodplain, the same power transmission corridor, or the same regional fuel supply. Even within one building, separate network rooms can share the same cooling system. Heat is not merely an inconvenience; it can be a failure mechanism. Electronics have thermal limits, batteries age faster in warmth, and cooling systems are themselves complex. If cooling fails, "redundant" servers can fail in a synchronized way because they are all subject to the same rising ambient temperature.

Another correlated dependency is software monoculture. Two separate routers made by the same vendor, running the same firmware, can fail together when a particular bug is triggered by an unusual packet or by a rare timing condition. When organizations standardize on one platform to simplify operations, they often increase the blast radius of a single defect. Standardization is rational. The risk is that it concentrates failure into fewer categories.

A subtler correlated dependency is *the failover mechanism.* Imagine a service that normally runs in one region and fails over to another during trouble. On paper, that is resilient. In practice, failover is not a gentle drift. It is a surge. The surviving region suddenly takes on extra load, extra cache misses, extra database writes, extra user sessions. The traffic spike coincides with a moment when the system is already under stress. If the surviving region was sized for average load plus a margin, the failover may shove it past a performance cliff.

This is where redundancy and overload meet. Systems often fail not because they run out of compute in a calm way, but because saturation triggers pathological behavior: queues grow, timeouts fire, clients retry, retries amplify load, and the system begins spending more effort on failing work than on successful work. A backup that is activated during a crisis inherits not only the legitimate traffic but also the storm of retries generated by clients experiencing errors.

In that moment, "more capacity" can behave like "more surface area for fire." The backup is not a spare tire; it is a second engine that must spin up at highway speed while the car is already skidding.

Chokepoints also arise from the economics of interconnection. The Internet is a network of networks, stitched together by contracts. A network that provides transit for many others becomes a de facto hub. Even if there are alternate transit providers, switching is not instantaneous; it can involve BGP convergence, policy changes, and sometimes manual coordination. Under pressure, networks tend to do what they have already practiced, which means that the usual paths become even more dominant during incidents. The "backup" path might exist but be poorly monitored, under-provisioned, or configured differently. When a sudden shift pushes traffic onto it, latent problems appear.

That is the paradox: redundancy can uncover the difference between a path that exists and a path that is truly ready.

All of this can feel abstract until it touches ordinary life. The reason chokepoints matter is not that engineers dislike them. It is that they quietly govern the boundaries of modern normality. When naming fails, "the Internet is down" becomes a lived experience. Work stalls. Logistics dashboards go blank. Payment terminals revert to offline modes. People cannot reach public services. Families cannot contact each other through familiar channels. In a society that increasingly uses the Internet as a default transport for speech, commerce, and coordination, an outage is not an inconvenience; it is a temporary reshaping of what is socially possible.

Chokepoints also matter because they can be attacked. An attacker does not need to destroy the whole system. They can aim at leverage points. In physical infrastructure, the attacker's challenge is access. In digital infrastructure, access can be substituted with volume. A denial-of-service attack is, in effect, a way to rent a crowd. With enough compromised devices, the attacker can place overwhelming demand on a chokepoint without ever touching it. When the chokepoint is a shared dependency like DNS, the attacker can affect many targets at once. The targets may never have heard of the chokepoint by name, yet their fate is tied to it.

It would be comforting to say that the answer is simply "make everything a mesh." But meshes, too, have centers, because humans and markets create centers. Even in a mesh, some nodes will be better connected, cheaper, or more trusted. Those nodes will attract dependence. They will become hubs in practice, even if not in topology. A pure mesh is also expensive, and expense is a kind of gravity that pulls systems back toward concentration.

So resilience becomes less about abolishing chokepoints and more about understanding their behavior under stress. That understanding has a few practical implications.

One is to treat redundancy as a *system*, not a pair of components. The question is not, "Do we have two?" It is, "Can we keep two consistent, secure, and testable?" A redundant setup that nobody rehearses will surprise you at the worst possible time. Failover should be exercised, not merely documented. The exercise should include the ugly parts: degraded networks, partial DNS failures, expired certificates, credential outages, and the human procedures needed when automation fails.

Another implication is to seek *independence* in redundancy. Different providers, different regions, different power grids, different software stacks when feasible, and different operational pipelines. Independence is not free; it increases complexity. But it buys something that simple duplication does not: resistance to correlated failure.

A third implication is to reduce unnecessary synchronization. Systems that require tight, instantaneous agreement across components are brittle. If every component must know the same truth at the same time, then a temporary disagreement can become an outage. Looser coupling can mean designing for stale reads, eventual consistency, local autonomy, and graceful degradation. Those phrases can sound like technical doctrine, but the human meaning is simple: when part of the system is confused, the rest should still do something useful.

Graceful degradation is the opposite of the all-or-nothing failure that chokepoints tend to induce. It means that when the "nice" path is unavailable, a "good enough" path remains. A website might serve cached content rather than personalized pages. A payment system might queue transactions rather than rejecting them. A routing system might prefer stability over optimization during turbulence. These are design choices, but they are also social choices about what to preserve when the world becomes unreliable.

Finally, chokepoints should be made *visible*. Many organizations discover their dependencies only during incidents. Dependency mapping sounds bureaucratic until you realize that it is a form of self-knowledge. If a service relies on a particular DNS provider, certificate authority, time source, cloud region, transit network, or identity provider, that reliance should not be a surprise. Visibility also enables collective action. When many organizations depend on the same linchpin, the linchpin becomes part of public infrastructure in effect, even if it is privately owned. That is a governance reality as much as an engineering one.

The most unsettling part of the chokepoint paradox is that it cannot be solved once and for all. It is a recurring pattern in complex systems: success concentrates, concentration creates leverage, leverage creates fragility, and fragility is often revealed only by stress. The task is not to eliminate all narrow places, which is impossible, but to prevent narrow places from becoming traps. The Internet's hidden nervous system can tolerate astonishing damage when its failures remain local and its recovery remains calm. What it cannot tolerate for long is the quiet accumulation of dependence on a few small places that everyone assumes will always be there.

6.2 The Domino Effect

A power line snaps in heavy wind. A router accepts one bad announcement. A database shard slows down for reasons nobody can reproduce. If modern infrastructure behaved like the objects on a desk, the damage would stay local: one broken thing, one repair. Instead, a small break can reorganize the stresses of an entire system. Load does not disappear; it *moves*. The moment something fails, its work is silently reassigned to whatever remains. The reassignment is often automatic, fast, and mercilessly literal. Neighbors inherit the burden, and if they were running near their limits, the burden tips them over. A cascading failure is that tipping-over, repeated.

The uncomfortable detail is that cascades are not only about weakness. They are also about efficiency. Systems are designed to run "hot" because unused capacity is expensive. Airlines do not schedule planes to sit idle. Data centers do not buy servers to do nothing. Power grids do not keep half their transmission lines empty just in case. The margin between normal and overloaded is often thinner than anyone would admit in a press release. Under those conditions, a cascade is not an exotic disaster. It is what happens when a system that has been tuned for everyday performance is suddenly asked to absorb a shock.

Engineers use the word *load* in many contexts, but it always means a demand that must be carried somewhere: electrical current, network traffic, customer requests per second, trucks rerouted from a closed highway, patients diverted from a full emergency room. Load shifting is the first move in the domino effect. It sounds like a solution—"route around the damage"—and often it is. Yet it is also a way of concentrating strain. When the system has no slack, routing around is just moving the same weight onto fewer shoulders.

In a power grid, the mechanics are stark. Electricity is not like water in a pipe that you can easily steer with valves. It follows the physics of the network: the distribution of impedances, the phase relationships, the geometry of the lines. Operators can influence flows by switching equipment and adjusting generation, but in the seconds after a line trips, power redistributes according to the grid's electrical laws. The redistribution is immediate, and it can push other lines beyond their thermal or stability limits. Protective relays then trip those lines to prevent equipment damage. Each protective trip removes another pathway, forcing the remaining network to carry even more current. The cascade accelerates because each act of protection changes the conditions for everything else.

A minimalist sketch of the logic looks like this:

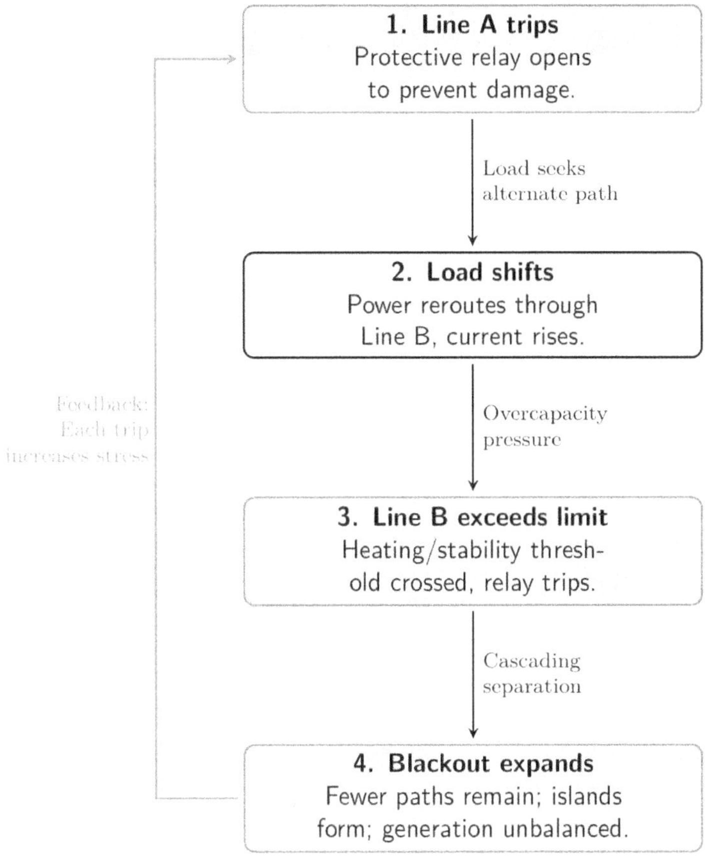

The diagram is deliberately simple, because real cascades are never polite enough to proceed in four neat steps. They branch. They pause. They restart. They unfold across time scales: seconds for relays, minutes for operators, hours for restoration, days for replacement equipment. But the essence holds: the first failure reshapes the network, the reshaping increases stress elsewhere, and protective actions can inadvertently act as accelerants.

A large blackout offers the most visceral illustration of the domino effect because it leaves physical traces: dark skylines, stalled elevators, broken supply chains, overheated apartments, and the uneasy quiet when traffic lights go out. One of the best-known cases in North America is the Northeast blackout of August 14, 2003, which affected a vast region of the United States and Canada. The initiating details included lines sagging into trees during hot conditions, alarms not handled as intended, and a system operating with less awareness than operators believed they had. The cascade itself was a chain reaction of line trips and rapidly shifting flows, producing a grid that split into islands that could not keep frequency stable. "One thing went wrong" is the wrong story. Many things were already close to their thresholds; one event made the whole arrangement reconfigure.

The same logic appears in the Internet, though it is easier to miss because the damage is often invisible. There is no darkness, only spinning loaders and vague error messages. Instead of current, the load is packets and requests. Instead of lines sagging into trees, the initiating event might be a small misconfiguration or a transient link failure. Instead of protective relays, there are routing protocols, traffic engineering policies, and automated failovers.

The Internet's domino effect often begins in the control plane: the set of signals that tell the data plane where to send traffic. BGP is a prominent example, but it is not the only one. A control plane failure is especially dangerous because it can cause healthy hardware to behave pathologically. When a router believes a destination is reachable through a certain path, it sends traffic there; if many routers believe the same wrong thing, huge volumes can converge on an unintended bottleneck.

A canonical Internet anecdote is the YouTube hijack of February 24, 2008. Pakistan Telecom announced a more specific route for a YouTube prefix in order to block the site domestically. A more specific prefix can override a legitimate broader route in the global routing system because many networks prefer specificity. When the announcement leaked to an upstream provider and propagated, large parts of the world began routing YouTube traffic into Pakistan, where it could not be delivered. The initiating act was small and local; the propagation was global.

What makes that episode a domino effect rather than merely a bad route is how quickly a distributed system can align on the wrong state. The Internet is decentralized, yet its participants use similar decision rules. Those common rules are a kind of coupling. If a rule causes everyone to prefer "the most specific route," then a mistaken most-specific route becomes a magnet. The traffic attraction can also create secondary failures: sudden congestion, overloaded links, and collateral damage to other services sharing the same paths.

A related but even more subtle domino pattern is the *retry storm*, sometimes called the *thundering herd.* The initiating event can be almost trivial: a backend service slows down slightly. Clients begin timing out. Many clients are written to retry automatically, because retries usually help: the network hiccup passes, the second attempt succeeds. Retries are a kind of everyday redundancy, a small act of resilience embedded in code.

But retries are also a feedback loop. When the system is merely slow, retries add more work exactly when the system is least able to handle it. They can turn a modest slowdown into a collapse. Worse, retries tend to synchronize. If a million clients time out around the same time, they may all retry around the same time, creating bursts that hammer the recovering service. The act of healing becomes another blow.

This is one reason large outages can feel paradoxical from the inside. Engineers may see that the system is "mostly working" and attempt a restart, a failover, or a capacity increase. Those interventions can temporarily worsen the situation, because each intervention triggers more retries, more cache misses, more reconnections, more authentication handshakes. The system has not only a failure mode but a *recovery mode*, and recovery can be the more dangerous phase.

The word "cascade" implies a one-way descent. In practice, cascades are often driven by loops. A failure causes load to shift; load causes overload; overload causes timeouts; timeouts cause retries; retries increase load; the increased load worsens overload. That is a positive feedback loop: an amplifier.

Positive feedback is not always bad. It is what allows a microphone to pick up a voice and make it audible to a crowd. It is what allows a thermostat to rapidly correct temperature when the heater is underpowered. The problem is that positive feedback without a brake produces runaway behavior. In infrastructure, the brakes are mechanisms that dampen demand and preserve partial function: load shedding, backoff algorithms, queue limits, circuit breakers, and admission control. When those brakes are absent, the system behaves like a crowd pushing through a doorway that is already jammed.

Load shedding deserves a plain-language translation because it can sound harsh. It means deliberately refusing some work so that the rest can succeed. A power grid sheds load by cutting off some customers to prevent a wider collapse. A web service sheds load by returning an error quickly rather than letting requests pile up and time out slowly. A cellular network sheds load by denying new calls during an emergency so that existing calls do not all drop. Load shedding is a controlled failure to avoid an uncontrolled one.

The emotional difficulty is obvious. It feels wrong to drop requests when the goal is to serve them. It feels wrong to cut power when the goal is to keep the lights on. Yet without load shedding, systems often fail in a way that serves nobody. A total blackout is worse than a controlled rolling outage. A service that times out for everyone is worse than a service that gives a prompt "try again later" to some users while continuing to serve others.

A domino effect can also jump between different kinds of infrastructure. Electricity and communications are tightly interdependent. Power stations and substations rely on communication networks for monitoring and control; communication networks rely on power for everything from cell towers to data centers. Those dependencies are often buffered by batteries and generators, but buffers have time limits. When a blackout is prolonged, communications begin to degrade. When communications degrade, the grid becomes harder to operate, and restoration slows. In the language of interdependent network research, this is coupling between layers: failures in one network disable nodes in the other, which in turn disables more nodes in the first.

That coupling helps explain why some failures look strangely abrupt. A system can appear stable and then, after one more small shock, collapse in a way that feels disproportionate. The system was not gradually weakening in an obvious way; it was drifting toward a threshold. Near that threshold, a small additional loss can trigger a large reconfiguration because the remaining network fragments. Buldyrev and colleagues (2010) showed mathematically that coupled networks can exhibit sudden, catastrophic transitions, unlike some single networks where damage accumulates more smoothly. The unsettling implication is that "it has held so far" does not guarantee "it will hold a little longer" when interdependencies are tight.

At a human scale, this threshold behavior shows up as surprise. A city copes with a heatwave until it doesn't. An airline network copes with a storm until one airport closure ripples into cancellations nationwide. A cloud platform copes with a regional glitch until a surge of automated failovers concentrates demand somewhere else. Surprise is often a sign that the system's true limits were not visible in normal operation.

The domino effect also exploits a weakness in how people assign blame. It is comforting to identify a single culprit: the tree branch, the operator, the bug, the misconfigured router. Yet cascades are usually overdetermined. Many preconditions align: tight capacity, imperfect monitoring, common-mode dependencies, automation without guardrails, incentives that reward efficiency over slack. The initiating failure is often replaceable; if it had not happened, another small disturbance might have served as the trigger next week.

That does not mean responsibility dissolves. It means responsibility shifts from the hunt for a single mistake to the craft of building systems that fail in a bounded way. Bounding failure is an architectural habit. It is also a social habit: practicing incident response, running drills, investing in observability, and admitting that "rare" events happen regularly when the world is large enough.

A striking feature of cascades in digital systems is that the boundary between "incident" and "everyday life" has blurred. Many services now run continuous experiments, continuous deployments, continuous optimization. The system is in motion even on calm days. Motion brings agility, but it also means there are more opportunities for small errors to be amplified by automation. A single bad configuration pushed quickly can become global before anyone has time to notice. The dominoes are close together, and the push travels at machine speed.

The most valuable mental shift is to stop imagining failure as a thing that happens to an isolated component, and to start seeing failure as a *redistribution problem*. When Line A trips, where does its load go? When a route is withdrawn, where does the traffic reconverge? When a region fails over, which shared databases become hot spots? When a resolver cannot answer, which caches expire and which clients stampede? Once you look for load shifting, you begin to see why cascades can be diagnosed too late: the first failure is not always the most damaging event. The damage comes from the system's reaction.

The hidden nervous system of civilization is not fragile because it is weak. It is fragile because it is tightly coupled, continuously loaded, and fast to reconfigure. Under those conditions, a small error can be the first domino not because it is powerful, but because everything nearby is already leaning. The moment the system begins to tip, the forces that normally keep it efficient—automatic rerouting, protective shutdowns, retries, failovers—can become the very mechanisms that carry the fall forward.

6.3 The Rising Tide

A great deal of modern infrastructure was designed with an unspoken promise: tomorrow will resemble yesterday closely enough that the past can be used as a safe guide. Engineers have a term for this comfort. *Stationarity* is the assumption that the statistical character of the environment—how hot it gets, how often rivers flood, how high storm surges reach—does not change in ways that matter over the lifetime of the asset. It is an assumption built into design codes, maintenance schedules, and insurance models. It is also an assumption that a warming climate is steadily withdrawing.

The collision is not abstract. The hidden Internet sits on top of physical things that expand, sag, corrode, swell, and drown. Fiber optic cables are hair-thin glass threads, but they depend on shoreline facilities, conduits, power feeds, and cooling systems. Cellular networks are "wireless" only in the last hop; behind every tower is a backhaul link, often fiber, and behind that is a chain of huts, cabinets, and switching centers. Data centers feel like pure software, yet they are among the most physical buildings in the modern economy, filled with copper busbars, diesel tanks, chillers, pumps, and miles of cable trays. Electricity is the first dependency for almost everything else, and the grid's own weakest moments often coincide with the very weather that is becoming more common: heat, drought, wildfire, and intense storms.

Climate risk is often described as a question of sea level, and sea level rise is indeed a slow-motion threat to coastal infrastructure. But the more immediate collision tends to come from extremes: days that are hotter than the equipment's derating curves assumed, storms that rain harder than drainage systems were sized for, wildfire seasons that last longer than a utility's vegetation management plan, and combinations of hazards that were once treated as independent. When hazards combine, the domino effect gains a new push.

Begin with heat, because heat is the simplest physical aggressor and the easiest to underestimate. Electricity transmission lines have a thermal limit: as current increases, the conductor heats; as it heats, it expands; as it expands, it sags. Sag is not cosmetic. A sagging line can violate safety clearances and strike vegetation. On a hot, still day, the line begins closer to its limit. At the same time, demand is often higher because millions of air conditioners are doing their best against the same heat. The grid is asked to deliver more power while the conductors can safely carry less. The margin narrows from both sides.

A similar story holds inside the Internet's physical plant. Electrical resistance in conductors rises with temperature. Electronics become more error-prone as junction temperatures climb. Fans spin faster, bearings wear, and filters clog. Cooling systems face a cruel arithmetic: the hotter the ambient air, the harder it is to reject heat, and the more energy must be spent on cooling rather than computing.

Data centers are designed with redundancy in chillers and power supplies, but redundancy does not repeal physics. If the environment exceeds design assumptions, the redundant components can be simultaneously stressed.

The fragility is not always in the computers. It is often in the interfaces: the point where "IT" meets "facilities," where software expectations meet mechanical realities. When cooling capacity is constrained, operators may throttle servers, which can trigger traffic shifts to other facilities. Those shifts can overload networks, or cause caches to miss, or increase latency just enough to trip timeouts, which can induce retries. A heatwave can therefore act like a systems test imposed by the atmosphere: it pushes multiple layers toward their limits at once.

Wildfire adds a different kind of stress: not only heat, but smoke, ash, and the deliberate de-energizing of power lines to prevent sparks. Smoke can foul cooling coils and filters. Ash can create conductive paths where none are wanted. And when utilities cut power as a safety measure, they are performing load shedding on a geographic scale. Battery backups and generators can bridge an outage, but the bridge has a length. Generators need fuel. Fuel delivery depends on roads, drivers, and a supply chain that may itself be disrupted by the same disaster. During long events, what fails is often not the prime system but the logistics that sustain the backup.

Hurricanes and intense storms add water to the mix, and water is an unusually versatile adversary. It can short-circuit electronics, undermine foundations, block access roads, saturate soil so that poles fall more easily, and flood basements that were once considered safely above the high-water mark. It can also knock out power and communications simultaneously, which makes coordination harder precisely when it is most needed.

Hurricane Sandy in 2012 is a memorable American example, not because it was the first severe storm, but because it revealed how easily "protected" infrastructure can become vulnerable when assumptions about storm surge and power resilience are exceeded. Flooded substations, dark neighborhoods, and fuel shortages turned backup plans into a scramble. Telecommunications infrastructure in lower Manhattan suffered outages, and even where network equipment survived, prolonged power issues and access constraints complicated restoration. Sandy's lesson was not merely "coastal cities are exposed." It was that infrastructures that normally act as each other's backups can fail together under environmental stress: the grid fails, generators depend on fuel, fuel depends on transportation, transportation depends on signals and communications, communications depend on the grid. The loops tighten.

Sea level rise makes those loops more likely to be triggered because it raises the baseline on which storm surge rides. The risk is not only that a facility is permanently inundated. The more common pathway is that rare events become less rare. A "once-in-a-century" floodplain becomes a "once-in-a-decade" headache. The financial and operational consequences are disproportionate, because critical facilities are not easily moved, and even when they can be rebuilt, they cannot simply be rebuilt everywhere at once.

Coastal vulnerability has a special relevance for the Internet because of undersea cables. The global Internet is stitched together by hundreds of submarine fiber optic cables. These are engineering triumphs, built to survive pressure, corrosion, and fishing activity. Yet every cable must come ashore somewhere, and where it comes ashore it depends on a cable landing station: a physical site with power, security, and backhaul connectivity into terrestrial fiber networks. Those landing sites are frequently near coastlines by necessity. Even if the cable itself is safe offshore, the onshore interface can be exposed to flooding and storm surge. A landing station is a chokepoint in a literal geographic sense: a narrow place where global connectivity meets local vulnerability.

A common misconception is that the Internet is "in the cloud," hovering above weather. The cloud is real; it is just made of buildings. Cloud concentration intensifies climate risk because it creates correlated exposure. A single metropolitan region can contain multiple data centers owned by different companies but supplied by the same grid, cooled by the same water system, served by the same transportation corridors, and subject to the same heat dome or smoke plume. Even when there are multiple regions, they can be linked by automated traffic management that shifts load quickly. Fast shifting is excellent for small faults; it can be dangerous under broad-area weather, because broad-area weather produces simultaneous strain across many locations.

Consider what happens during a regional heatwave in a major demand center. The grid operator may issue conservation alerts. Some industrial loads may be curtailed. Rolling outages may be used to protect the stability of the grid. At the same time, cellular networks face increased load as people seek information, check on family, or stream news. Home Wi-Fi routers and modems may be power-cycled by brief outages, leading to reconnection storms when power returns. Data centers may switch to generator power, which changes their internal risk profile. Traffic engineering systems may reroute traffic away from affected facilities, increasing load elsewhere. Each of these responses is rational in isolation. Together, they can create the conditions for a cascade in the digital layer, triggered by a physical stress that is becoming more common.

This is where the earlier discussion of chokepoints and cascades becomes more than metaphor. Climate change is not just a new category of "bad days." It changes the distribution of stress in time and space. It increases the frequency of compound events: heat plus wildfire smoke; storm surge plus inland flooding; drought plus high demand. Compound events are especially good at breaking redundancy, because redundancy often assumes that failures are independent. If two "independent" sites share a river basin, a power market, a wildfire corridor, or a coastal plain, the environment can synchronize their stress.

Some of the most severe climate-related impacts on infrastructure are also the least cinematic. Drought can reduce hydroelectric generation and limit the cooling water available for thermal power plants. Low river levels can restrict barge traffic that delivers fuel. High water temperatures can force power plants to reduce output because discharge water would be too warm for environmental permits. Each constraint reduces available generation precisely when air-conditioning demand spikes. The grid becomes tighter; the probability of load shedding increases. Digital infrastructure then inherits the consequences of a grid that is operating closer to its edge.

Even when the Internet's own equipment remains physically intact, climate stress can change the economics of maintenance and the tempo of operations. Hotter conditions accelerate wear on batteries. Batteries are deceptively central: they bridge the milliseconds between grid power loss and generator stabilization; they buffer transient events that would otherwise reboot equipment. As batteries age faster, the probability of a "brief" power disturbance causing a larger outage rises. Operators can compensate with more testing, more replacement, more monitoring. That is cost. When budgets are tight, deferred maintenance becomes a hidden form of climate vulnerability.

There is also a social dimension. Climate-related disasters do not simply break equipment; they thin the human capacity to respond. During extreme events, staff may be unable to reach sites. Roads may be closed. Evacuation orders may limit access. Employees may be coping with their own emergencies: flooded homes, sick relatives, heat stress. A recovery plan that assumes the calm availability of skilled labor can fail under conditions that are simultaneously technical and humane.

The Internet's governance layer intersects with climate risk in unexpected ways. Routing security practices like RPKI origin validation and operational initiatives like MANRS are usually discussed as defenses against malicious hijacks or sloppy route leaks. Yet they also act as resilience measures in a world where emergencies are more frequent. When networks are stressed by weather and traffic shifts, the last thing the system needs is a control-plane accident that misdirects traffic into a bottleneck. A routing ecosystem with better hygiene reduces the chance that, during a wildfire evacuation or a hurricane landfall, connectivity will be further undermined by preventable misconfiguration.

The same is true of denial-of-service capacity. The growth of poorly secured Internet-connected devices expands the pool of machines that can be conscripted into botnets. In a warming world, disruptive events create attention, and attention creates opportunity: attackers know that defenders are busy and that degraded conditions make services more brittle. A DDoS attack against a DNS provider or a regional ISP during a climate emergency can have outsized impact, not because the attack is novel, but because the system has less slack to absorb it.

There is a temptation to treat climate adaptation as a matter of hardening: build higher seawalls, install bigger generators, add more cooling. Those measures help, but they can also reinforce a fragile assumption: that the future environment can be met by scaling up the same design, rather than by changing how systems degrade and recover. Bigger walls and bigger generators do not necessarily address the coupling problem. If everything depends on continuous high-quality power, then the system remains vulnerable to any prolonged fuel disruption. If everything depends on continuous perfect connectivity, then the system remains vulnerable to any outage that triggers retry storms. Hardening without rethinking can lead to what might be called resilience theater: impressive investments that still fail in the same systemic ways.

Adaptation, in practice, often looks less heroic and more disciplined. It means treating heat as a primary design constraint rather than an exception. It means siting critical facilities with an eye toward future flood maps, not just historical ones. It means diversifying supply chains for spare parts and fuel, and ensuring that access routes exist under disaster conditions. It means designing services that can run in a degraded mode for longer than a few minutes, because climate-driven disruptions may not be brief.

It also means revisiting the cultural habit of assuming that rare events will remain rare. The language of "hundred-year flood" has always been misleading; it describes a probability, not a schedule. In a changing climate, the probability itself changes. Infrastructure planning that treats yesterday's probabilities as permanent is planning to be surprised.

None of this implies that the nervous system of civilization is doomed to fail more often. It implies that the risk landscape is moving, and that a system optimized for a stationary world will feel increasingly brittle. The most important shift may be psychological: to stop thinking of weather as background and start thinking of it as an active participant in network behavior. A heatwave is not merely uncomfortable; it is a load test for power transmission, cooling systems, batteries, and human response capacity. A storm surge is not merely a coastal event; it is a threat to the handful of physical sites where global connectivity meets the shore. Smoke is not merely an air quality issue; it is a particulate stressor for the fans and filters that keep computation alive.

The rising tide is not only literal water. It is the rising baseline of stress on systems that were built with narrower margins, sharper couplings, and stronger assumptions of normalcy than the future is likely to permit. When that stress arrives, failure is rarely a single spectacular break. It is more often a sequence of small, sensible responses that, taken together, push the system across a threshold where it can no longer easily return to balance.

Chapter 7

Shadows in the Network

The internet's founding protocols rely on a naive trust that assumes all nodes act in good faith, a design choice that has become a critical liability. We investigate the mechanisms of modern digital warfare, where the logic of routing is subverted to redraw the map of the world without moving a physical border.

7.1 Weaponized Packets

A denial-of-service attack is, at heart, a crude idea: if you cannot break into a building, block the doors until nobody can get in. On the internet, the "doors" are a service's limited capacity to receive and answer messages. That capacity is never infinite. Servers have finite CPU cycles, memory, and network links; upstream providers have finite bandwidth; firewalls have finite tables for keeping track of connections; help desks have finite patience. Distributed Denial of Service (DDoS) attacks take that basic scarcity and industrialize it, turning the internet's own habits—its defaults, its conveniences, its optimism about good behavior—into a weapon.

The unsettling part is not that a determined adversary can generate a lot of traffic. Many organizations can generate a lot of traffic. The unsettling part is the *asymmetry*: an attacker can spend very little to force a defender to spend a lot, and can do so from a distance with a degree of deniability that would be impossible in the physical world. A stone through a shop window costs the attacker a stone; a DDoS attack can be a pebble that triggers an avalanche.

To see why, it helps to think in packets. A packet is a small parcel of data, wrapped with addressing information so it can travel across networks. The internet's routers treat packets with a kind of bureaucratic indifference: forward this packet toward its destination, then move on to the next. This indifference is a strength—it keeps the system fast and scalable—but it also means that much of the internet does not ask deep questions about *intent.* A router does not know whether a packet is a heartfelt email, a software update, or a punch thrown in the dark. It often cannot know, and historically it has not been required to.

Many DDoS attacks take advantage of that indifference by overwhelming the most expensive part of the chain. The defender's costs are often non-linear: once a link is saturated, legitimate packets are dropped; once a firewall's state table fills, new connections fail; once an application's worker threads are busy, requests queue; once queues fill, they spill into timeouts. Meanwhile, the attacker's costs can be linear and small: send more packets, or better yet, send *strategic* packets that induce others to send packets on your behalf.

The simplest form of DDoS is volumetric flooding: a large number of machines send traffic directly to the victim. If enough traffic arrives, the victim's internet connection becomes congested, like a highway filled with cars that have no destination other than the traffic jam itself. But direct flooding has a limitation: the attacker must supply the volume. As defenders improved their capacity and learned to buy protection from specialized networks, attackers learned a trick that looks like a magic act until you notice the trapdoor: *reflection and amplification.*

Reflection uses a third party as an unwitting mirror. Amplification uses the mirror to make the reflection larger than the original. The stage prop is often UDP, the User Datagram Protocol. UDP is a lightweight way to send a message without first establishing a connection. Unlike TCP (the Transmission Control Protocol), which performs a handshake before data flows, UDP is happy to fire-and-forget. It is useful for streaming, real-time games, voice calls, and a host of infrastructure services. It is also useful to attackers because many UDP-based services reply to a request without verifying much about who sent it.

If an attacker can *spoof* the source address on a UDP packet—that is, lie about where it came from—the reply will be sent to the victim instead. The attacker sends a small request to a public server, forging the victim's address as the sender. The public server, doing what it was designed to do, sends a response to the victim. Multiply this across thousands of public servers and the attacker has turned the internet into a choir of strangers shouting at the target on command.

Amplification enters when the response is larger than the request. Some protocols, through no malice and plenty of historical accident, can be coaxed into sending far more data back than they received. The attacker pays for the postage on a postcard; the victim receives a crate. The ratio between response size and request size is the *amplification factor*. Even a modest amplification factor matters when multiplied at scale. An attacker's upstream link might only support a small volume; reflected, amplified traffic can produce a flood far beyond that original constraint.

The economics here are as important as the mechanics. DDoS is a business model as much as an attack technique. It thrives when the cost per unit of harm is low and the cost of defense is high. A defender can buy more bandwidth, but that is an ongoing expense. A defender can deploy scrubbing services, but that is a recurring contract and an operational dependency. A defender can redesign software to degrade gracefully, but that is engineering time, complexity, and tradeoffs. An attacker can rent a botnet by the hour, or purchase access to "booter" services that advertise DDoS on demand, and shift targets when defenders harden. The attacker can fail repeatedly at low cost; the defender must succeed continuously at higher cost.

The reflection trick also fits the internet's sociology. The attack traffic comes from legitimate machines and legitimate services. The victim sees a storm of packets from innocent servers scattered across the world, each of which thinks it is replying to a request. Blame becomes slippery. Who attacked you: the spoofing machine, the reflecting server, the ISP that allowed spoofing, the manufacturer that shipped insecure devices, or the market that rewarded speed over security? The answer is, in some sense, "yes."

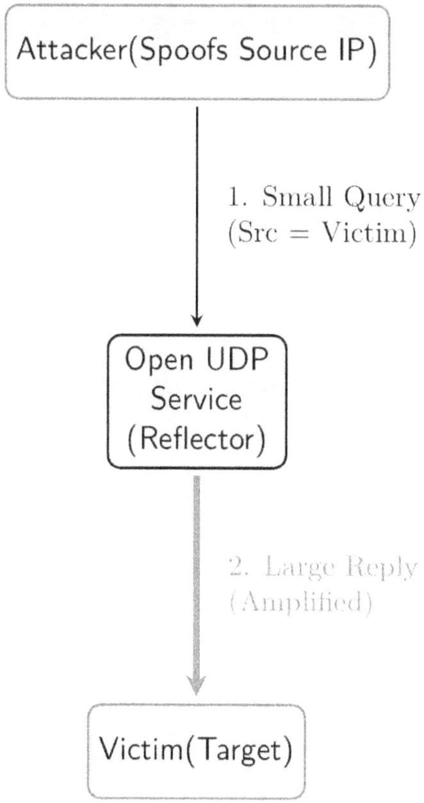

$$\text{Amplification Factor } A \approx \frac{\text{Bytes Out}}{\text{Bytes In}}$$
$$\textit{Victim Load} \approx A \times \textit{ Attacker Cost}$$

This diagram hides an important reality: the reflector is rarely a single server. In a real attack, reflectors are many and diverse: misconfigured DNS resolvers, exposed NTP servers, public memcached instances, gaming servers, and other UDP-speaking systems that were never meant to be part of a conflict. Each reflector may contribute only a trickle. Together, they become a river.

Spoofing, the essential lie at the center of reflection, depends on a failure of policing at the edges of the network. In principle, an ISP can prevent customers from sending packets with forged source addresses. In practice, adoption has been uneven, especially in parts of the internet where cost pressures are intense and accountability is diffuse. The result is that the internet still contains regions where a machine can claim to be any address it likes, at least for long enough to cause trouble.

Amplification is the other half, and it is as much about protocol design as about misbehavior. Take DNS, the Domain Name System, which maps names to addresses. DNS is foundational; it is also chatty. A single question can elicit a long answer, especially if the question is crafted to request a large set of records. DNS was built in an era where the adversary model was closer to "occasional misconfiguration" than "organized abuse at scale." Over time, extensions were added, making responses larger, and many servers were left open to queries from anywhere. A similar story exists for NTP, the Network Time Protocol: useful, ubiquitous, and in some configurations willing to answer a small query with a much larger reply. Memcached, a caching system used to speed up websites, became an amplifier when exposed to the public internet with UDP enabled; a tiny request could trigger a very large response, and for a time it became a notorious component in record-setting attacks.

Even when amplification factors are not enormous, DDoS can still be devastating because the defender is not merely trying to absorb bytes. Many services are shaped by hidden bottlenecks: the maximum number of connections a load balancer can track, the rate at which a database can open new sessions, the capacity of a logging system that writes to disk. Attackers learn to aim at the bottleneck that is hardest to scale quickly. Sometimes that bottleneck is not the victim at all, but a dependency: an authentication provider, a DNS host, an API gateway, a payment processor. Modern online services are less like castles and more like crowded city blocks of interconnected businesses; blocking one street can gridlock the neighborhood.

The human experience of a DDoS attack is oddly intimate. For users it feels like a door that refuses to open: pages hang, video buffers, logins fail. For operators it is a stressful flood of dashboards turning red, alerts piling up, and uncertainty about what is failing first. A congested link looks like packet loss and latency spikes. An overwhelmed application looks like timeouts and error codes. Under pressure, the instinct is to make changes quickly—restart servers, scale up instances, add rules to firewalls—but the attack itself is a moving target. DDoS is often designed to induce mistakes: to force the defender to block legitimate traffic, to exhaust on-call teams, to create a sense of helplessness that is itself a form of harm.

The scale of this harm became widely visible in the mid-2010s as botnets evolved from hijacked PCs into something more unsettling: swarms of everyday devices. In 2016, the Mirai botnet showed how quickly insecurity in consumer "Internet of Things" devices could be converted into raw force. Mirai spread by scanning the internet for devices such as cameras and routers that still used default passwords or weak, hard-coded credentials. Once compromised, devices could be instructed to send traffic at a target, turning cheap hardware into a hired mob. The deeper problem was not that a particular piece of malware existed; malware always exists. The deeper problem was that millions of devices were designed and shipped with the assumption that they would not be attacked, and that the consequences of their compromise would be limited to the owner. Mirai inverted that assumption: a device's insecurity became a public nuisance, a kind of environmental pollution in packet form.

There is a grim neatness to the Mirai story as an anecdote about modern risk. The owner of a compromised camera might never notice. The camera might still record and stream as expected. But the camera also became a participant in a siege against someone else. This is weaponization by indirection: the victim is attacked not only by the attacker, but by the attacker's ability to draft other people's property into the conflict. It is as if a vandal could, without leaving home, cause every lawn sprinkler in a neighborhood to turn toward a single house.

Reflection attacks draft infrastructure rather than devices: servers and services that were built to be helpful end up being coerced into mischief. Botnets draft endpoints rather than infrastructure: devices that were built to serve their owners become soldiers. Both forms of drafting share a moral hazard. The cost of poor security is often externalized. Manufacturers rush products to market; consumers value low price and convenience; ISPs optimize for connectivity; and the aggregate result is a pool of vulnerable machines that attackers can exploit. The market that produces the internet's abundance also produces the internet's surplus of exploitable capacity.

Not all DDoS attacks are about raw bandwidth. Some are about *state*, the memory a system uses to keep track of ongoing conversations. TCP, for example, requires both sides to agree to talk; that agreement is a handshake. A classic attack is to send many handshake initiations but never complete them, leaving the server holding half-open connections. Even when mitigations exist, the pattern remains: force the defender to allocate scarce internal resources for interactions that will never become useful. Another pattern is application-layer DDoS: instead of sending many packets, send carefully crafted requests that are expensive to handle—a search query that triggers complex database work, or a page that forces the server to render a large template, or a request that thrashes a cache. Here the attacker's packets are few, but each one forces the defender to do disproportionate work. The siege is no longer at the gates; it is inside the kitchen, demanding elaborate meals and discarding them untouched.

Defenders respond with strategies that sound almost like urban planning. One strategy is to widen roads: buy more bandwidth, add more servers, distribute services geographically. Another is to build checkpoints: rate limiting, filtering, reputation systems, and the careful distinction between "normal bursts" and "abuse." Another is to outsource policing to specialists: DDoS scrubbing centers that can absorb large floods and forward only clean traffic. Another is to change the rules of engagement: require cryptographic puzzles or tokens before expensive work is done, so that the attacker must pay a computational cost per request.

But every defense has a shadow cost. Filtering can block legitimate users, especially those behind shared networks. Rate limiting can break applications that rely on bursts. Outsourcing can centralize power and create new single points of failure. Cryptographic puzzles can disadvantage low-power devices and increase energy consumption. Even "just add capacity" has a limit: there is always a bigger flood, and the defender's marginal cost per extra unit of capacity is often higher than the attacker's marginal cost to rent or coerce more sources.

A particularly revealing defense is *anycast*, a technique where the same IP address is advertised from many locations around the world. Requests to that address are routed to the nearest or "best" location according to the internet's routing decisions. Anycast is widely used for DNS and content delivery. Under attack, anycast can spread incoming traffic across many sites, preventing any single location from being overwhelmed. In the language of sieges, it is as if the city could suddenly grow many gates and teleport arriving crowds to whichever gate is emptiest. Yet even anycast has constraints: it can complicate troubleshooting, and routing decisions can shift unpredictably under load. The internet is not a single machine; it is a federation of networks with their own incentives. Defenses must work with that messy reality.

DDoS also has a political dimension. A denial-of-service attack is often dismissed as a childish prank: loud but not clever, disruptive but not transformative. That underestimates its utility as a tool of coercion. DDoS can silence a newsroom during an election, disrupt a financial service during market hours, or harass an activist organization until its volunteers burn out. It can function as extortion: pay or stay offline. It can act as cover for other actions, distracting defenders while data is stolen elsewhere. It can also serve as a demonstration of power, a way of saying, "We can reach you whenever we want." Even when the technical harm is temporary, the psychological effect can linger.

The asymmetry becomes clearest when you consider how a DDoS attack chooses *where* to impose cost. If the attack saturates a link, the victim's upstream provider must deal with it. If the attack overwhelms a DNS provider, the victim may vanish from the internet's map even if its servers are healthy. If the attack targets a small organization's website, the organization may not have the resources to buy industrial-grade protection, and the site becomes collateral in a world where attention is concentrated on the largest platforms. The internet's nervous system carries signals for everyone, but not everyone can afford equal protection for their synapses.

There is a quiet lesson here about design choices that seemed reasonable when the internet was smaller and more collegial. Protocols that answer questions without asking who asked them are convenient.

Networks that prioritize forwarding over inspection are efficient. Devices that ship with simple setup procedures are easy to sell. Each choice, viewed locally, is defensible. The harm appears only when millions of such choices interlock and an adversary learns to play them like an instrument.

Weaponized packets do not require genius. They require leverage. Leverage comes from the gap between what a system assumes and what a system will tolerate. The internet assumed that most participants would not lie about their addresses, that most public services would not be abused at global scale, that most consumer devices would not be permanently compromised, and that most conflicts would not be fought through infrastructure shared by everyone. DDoS attacks are the art of exploiting those assumptions. When they succeed, the victim's failure is public and immediate, while the attacker can remain hidden behind the very abundance that makes the internet useful.

What makes this form of violence distinctive is its impersonality. A DDoS attack does not have to hate you. It does not have to understand you. It merely needs your service to be reachable, and therefore limited, and therefore vulnerable to being crowded out. The packet does not carry anger; it carries indifference, multiplied until indifference becomes force.

7.2 The Theft of Geography

Most people picture the internet as placeless: you tap a screen in Chicago and a page appears as if from the air. Yet the internet is obsessed with geography. Not the geography of mountains and rivers, but a geography of ownership: which network is willing to carry whose traffic, and on what terms. Those decisions are made constantly, automatically, and mostly invisibly. When they go wrong—

through accident, negligence, or sabotage—the internet does something uncanny. It redraws its own map.

To understand how that can happen, it helps to separate two questions that everyday browsing fuses into one. First: *What is the name of the thing I want?* That is the job of DNS, the Domain Name System. Second: *Where, in numerical terms, is it reachable?* That becomes an IP address. Third: *Which path through the world's networks should my packets take to reach that address?* That is the job of BGP, the Border Gateway Protocol. DNS answers the "who," BGP answers the "how," and together they decide whether your request reaches the intended destination or takes a detour through someone else's hands.

BGP is sometimes described as the internet's postal service for big institutions. Your laptop doesn't speak BGP. Your home router probably doesn't either. But the networks run by internet service providers, cloud companies, universities, and national backbones do. They speak to each other in a language of reachability: "If you want to deliver packets to this range of addresses, send them to me." Those ranges are called *prefixes*. A prefix is like a neighborhood in the address space. An announcement of a prefix is like putting a sign on the highway: "Downtown this way."

The striking feature of BGP is how much it relies on trust and convention rather than hard verification. BGP was designed for a world where the number of major networks was manageable and where misbehavior was assumed to be rare. Its primary goal was to let independently run networks coordinate without a central authority. That independence is one of the internet's great strengths. It is also a source of fragility: a system that must work for everyone can be difficult to secure against the subset of participants who choose not to play fairly.

An internet "path" in BGP terms is not a list of cities but a list of *autonomous systems* (ASes). An autonomous system is, roughly, a network under one administrative control: a large ISP, a cloud provider, a content delivery network, a university, a big enterprise. Each has an identifying number, an ASN. BGP announcements say not only "I can reach prefix *P*," but also "Here is the sequence of ASes this information passed through." That sequence is called the *AS_PATH*. It functions like a trace of custody, though not a cryptographically sealed one.

Two properties make BGP powerful and, in the wrong circumstances, dangerous. The first is that it is distributed. There is no single control room where routes are approved. Networks negotiate, announce, withdraw, and filter routes according to their policies and economic relationships. The second is that route selection follows simple rules. One of those rules is easy to state and surprisingly consequential: *more specific wins*. If one network announces a broad prefix and another announces a narrower prefix contained inside it, the narrower one is typically preferred. It is like a street sign that says "All mail for the city goes here," competing with another sign that says "All mail for this one block goes here." The block sign wins, because it looks more precise.

This is not a bug; it is a feature. It allows network operators to steer traffic during outages, balance load, and carve out sub-regions for engineering reasons. But it also creates an opportunity for hijacking. If an attacker, or simply a careless operator, announces a prefix that they do not legitimately control—especially a more specific one—they can attract traffic that was never meant to reach them.

A famous example occurred on February 24, 2008, when Pakistan Telecom attempted to block YouTube domestically. One technique for doing that is to announce a route to YouTube's addresses that points to a "black hole," causing traffic to disappear. The plan was local: make YouTube unreachable for customers in Pakistan. The mechanism, however, was global. The announcement leaked beyond its intended boundary, was propagated by an upstream provider, and spread across the internet. For a time, much of the world followed the "more specific" route and sent YouTube traffic toward Pakistan. A national censorship maneuver briefly became an international outage.

The details matter because they show how little malice is required. No Hollywood hacker needed. A single network made an announcement that should never have escaped; another network failed to filter it; the rest of the internet did what it is designed to do: share reachability information quickly. The very speed that makes the internet resilient to random failures makes it vulnerable to fast-moving mistakes.

Mistakes are not the only concern. Hijacking can be intentional, and it can be subtle. The crude version is a black hole: attract traffic and drop it, causing an outage. The more interesting version is a *traffic detour*: attract traffic, inspect or manipulate it, and then forward it on to the real destination so that the user may not notice. Detours are the internet's version of a roadblock that waves you through after taking down your license plate and photographing your passengers. It can enable espionage, credential theft, or content injection. Even when traffic is encrypted, detours can reveal metadata: who is talking to whom, when, and how much.

Here BGP and the physical world meet. A detour is not merely a logical event; it reroutes data through actual equipment in actual places. Packets are still carried by fiber-optic cables, undersea routes, metro rings, routers in data centers. But the *choice* of those physical paths is not based solely on distance. It is based on policy and economics: peering agreements, transit contracts, congestion, and the local decisions of network operators. Geography, in this sense, is created by business relationships. That is why it can be stolen.

One might ask: why not add authentication? Why not make BGP announcements cryptographically verifiable, so that only the rightful holder of a prefix can announce it? Over the years, the internet community has built partial answers, often under the umbrella of RPKI (Resource Public Key Infrastructure). The idea is straightforward: create a way for the holder of an IP prefix to publish a signed statement authorizing certain ASes to originate it. Those statements are called ROAs, Route Origin Authorizations. Routers can then perform *origin validation*: if AS X claims to originate prefix P, check whether X is authorized for P.

Origin validation helps, but it is not a complete solution. It validates only the origin, not the entire path. A route might originate from the correct AS and still be carried along an unexpected or malicious route. There are proposals for full path validation, such as BGPsec, which would add cryptographic signatures to attest to each hop of propagation. The difficulty is not that cryptography is unknown; it is that deployment at global scale is hard. Routers are optimized for speed, operators fear breaking connectivity, and incentives do not always align. Security is a shared benefit with individual costs, and the internet is a collection of independent actors.

Even without attackers, there is another class of failure that reveals BGP's reliance on tacit assumptions: the *route leak*. A route leak occurs when a network accidentally advertises routes to a neighbor that it should not. The classic case is a customer network that receives full internet routing information from a provider and then, by misconfiguration, announces those routes to another provider or peer, effectively claiming that it can provide transit for destinations it should not. The leaked routes may look perfectly ordinary to others. They may even offer an apparently shorter path. Traffic then flows through a network that is not prepared to carry it, causing congestion, outages, or unintended surveillance opportunities.

Route leaks expose an uncomfortable reality: BGP encodes business relationships implicitly, as unwritten rules. "Customers can send routes to providers, but providers should not send certain routes back to other providers via a customer," and so on. These norms exist because they make economic sense and keep traffic flowing efficiently. But norms are not protocols. When a configuration error violates a norm, BGP itself does not necessarily protest. It obeys, because obedience is its job.

Engineers have attempted to formalize these roles. One modern approach is for BGP sessions to include an explicit declaration of relationship roles (customer, provider, peer, and so on), so that certain classes of leaks can be prevented or detected. This helps turn social expectations into machine-checkable constraints. Yet the underlying theme remains: the internet's control plane is a patchwork of voluntary cooperation, and voluntary cooperation can fail.

If BGP can be used to steal geography by rerouting packets, DNS can be used to steal identity by rerouting *names*. DNS is the system that tells you, when you type a human-friendly domain name, which IP address to use. DNS is both mundane and sacred: if it lies, the rest of the security stack has to work harder.

DNS is distributed in a different way from BGP. Rather than a mesh of networks exchanging routes, DNS is a hierarchy of delegation. At the top are the root servers. Beneath them are top-level domain servers (for domains like `.com`, `.org`, `.net`, and national domains). Beneath those are authoritative servers for individual domains. Your computer usually doesn't talk directly to all these layers. It asks a *resolver*, typically run by your ISP, your company, or a public service. The resolver does the work of walking the hierarchy, caching answers, and returning them quickly.

DNS's original design assumed that answers could be trusted. That assumption was not irrational in the early internet; it has aged poorly. One classic attack is *DNS cache poisoning*, where an attacker tricks a resolver into storing a false answer. Once poisoned, the resolver will serve that wrong answer to many users, sending them to the attacker's chosen address. This can be done through clever guessing of transaction identifiers and timing, through exploitation of misconfigurations, or through compromise of the resolver itself. The key point is that DNS is an amplifier of trust: one bad answer cached in the right place becomes a bad answer for thousands or millions of people.

Like BGP, DNS has a security retrofit: DNSSEC, which allows DNS records to be signed so resolvers can verify their authenticity. DNSSEC makes poisoning much harder when deployed correctly. Yet deployment is uneven, and even with DNSSEC there are practical complications: key management, operational errors, and the reality that not all resolvers validate signatures. A system can be secure in theory and fragile in practice.

BGP hijacking and DNS poisoning share a family resemblance. Both exploit the fact that the internet's "map" is not a physical map but a set of claims. BGP claims "this prefix is reachable via me." DNS claims "this name corresponds to that address." When those claims can be falsified, the attacker does not need to break encryption or compromise servers; they can simply redirect the flow of traffic so that users knock on the wrong door.

This redirection has a particular menace because it attacks the *invisible* layer. People often focus on content: what a website says, what an email contains, what a message reveals. The internet's control systems are about reachability itself. They decide which messages arrive, and where. A redirection attack can be invisible to the user because the browser still displays a familiar name, the connection still opens, and the page still loads. The difference may be in the background: a warning that is ignored, a certificate prompt that looks technical, a slightly slower response time, a subtle change in behavior. Many users, and many organizations, are trained to treat connectivity as a given. Redirection attacks target that trust.

There is also a darker political use: the ability to withdraw routes. When a network stops announcing its prefixes via BGP, the rest of the internet no longer knows how to reach it. This is not smashing a cable; it is removing the destination from the map. During the Egyptian uprising in January 2011, connectivity was sharply curtailed in a way that, as contemporaneous reporting emphasized, appeared to involve major networks withdrawing routes. The effect resembled a drawbridge being raised: the physical infrastructure remained, but the global routing system behaved as though the country had become unreachable.

That kind of event reminds us that "internet freedom" is not only about what content is allowed. It is also about whether the paths exist at all, and who controls the announcements that make those paths visible. Because BGP is operated by networks, not by end users, the levers of connectivity can be pulled far upstream from where people experience the consequences.

The defender's predicament is a familiar one by now. Traffic can be encrypted, applications can be hardened, servers can be patched, and yet the control plane can be attacked in ways that shift the ground underfoot. Monitoring can help: projects like RouteViews and RIPE RIS collect BGP data from many vantage points, allowing researchers and operators to observe anomalies and reconstruct incidents after the fact. Detection systems can flag suspicious announcements or unexpected changes in paths. But detection is not the same as prevention, and prevention is not the same as rapid response. When a hijack occurs, the first battle is often to *notice* it, and the second is to persuade other networks to filter or correct it.

All of this creates a peculiar form of insecurity: your safety depends on other people's configurations. Your packets might traverse networks you have never heard of. Your domain might depend on resolvers you do not control. Your service might be reachable only as long as your upstream providers propagate your routes correctly and filter everyone else's. This is the hidden internet's social contract: a vast, largely cooperative system whose reliability emerges from the average competence and goodwill of many parties, and whose failures can be spectacular because the system is so interconnected.

There is no single villain here, and that is part of the discomfort. A route leak can be an honest mistake. A hijack can be a criminal act or a state operation. A DNS poisoning event can be a clever exploit or a sloppy compromise. What unites them is the possibility of *stealing geography* without moving a single cable: making New York look like it is in Karachi, making a bank's domain look like it lives on an attacker's server, making a country's networks look like they have vanished.

The modern internet works because it continuously answers two questions at high speed: "What do you mean?" and "Where is it?" When those answers can be forged, the internet's nervous system becomes susceptible to phantom limbs and false signals. A packet still travels somewhere; the tragedy is that "somewhere" may be chosen by whoever learned how to redraw the map.

7.3 The Mathematics of Privacy

Privacy used to be an architectural feature: a door, a curtain, a sealed envelope. Online, privacy is more often a claim about a hostile environment. Packets can be delayed, copied, redirected, or dropped. Names can be poisoned. Routes can be hijacked. Devices can be drafted into someone else's attack. In that setting, privacy stops being a matter of good intentions and becomes a matter of mathematics: proofs instead of promises, worst cases instead of average cases, adversaries instead of accidents.

Encryption is the simplest expression of that shift. It is often explained as "scrambling" data so only the intended recipient can read it. That is true as far as it goes, but it undersells what encryption really does for modern civilization. Encryption is a way to create islands of certainty inside a system that cannot be trusted. It does not make the network benevolent. It treats the network as if it might be watched, tampered with, or misdirected, and then asks: what can still be guaranteed?

The basic guarantee of encryption is confidentiality: without the key, an eavesdropper cannot feasibly recover the message. The word *feasibly* matters. Cryptography is not magic; it is engineering under assumptions about computation. Modern encryption depends on the belief that certain problems are hard enough that an attacker, even with large resources, cannot solve them within the lifetime of the secret. When those assumptions change, cryptography changes too. That is part of its austerity. It is an arms race, but one fought with proofs, reductions, and careful definitions.

A second guarantee is integrity: a recipient can detect if a message was altered in transit. This is the difference between a sealed envelope and a sealed envelope with tamper-evident wax. On a network where packets can be intercepted and modified, integrity is often as important as confidentiality. In the early days of the web, you might have visited a site in plaintext HTTP, assuming that "reading" was harmless. Yet if an intermediary could alter what you received, reading became a way of being infected. Integrity is what makes it possible to trust that a software update really came from the vendor, that a bank login page is not a forged imitation, that the words you see were not silently rewritten along the way.

Modern internet encryption, embodied by protocols like TLS (Transport Layer Security), aims to provide both confidentiality and integrity between your device and a server. In practice, this means that when you see a padlock icon, your browser has attempted to establish an encrypted channel, and part of that establishment involves authenticating the server. Authentication is where the mathematics of privacy meets the politics of governance, because the question "who is the server?" becomes "who gets to vouch for identity?"

The web's answer is the Public Key Infrastructure (PKI), a system of certificates and certificate authorities (CAs). A certificate is a signed statement that a particular public key belongs to a particular domain name. Certificate authorities are entities trusted by browsers and operating systems to issue those statements. In theory, this is a neat solution: many independent authorities, each able to attest to identities, with browsers checking signatures automatically. In practice, it creates a fragile apex of trust. If a CA is compromised or coerced, it can issue valid-looking certificates for domains it does not control, enabling man-in-the-middle attacks that break the intended guarantees of TLS without triggering browser warnings.

The history of PKI contains enough cautionary tales to make the point vivid. In 2011, the Dutch certificate authority DigiNotar was compromised, and fraudulent certificates were issued for high-profile domains. The incident was not merely a technical failure; it was a governance crisis. Browsers responded by removing trust in DigiNotar, an act that effectively destroyed the company's role in the ecosystem. The episode revealed an uncomfortable truth: the security of encrypted web browsing depends on the security practices of a long list of organizations that most users will never hear about, operating under a variety of legal regimes and incentives.

If that sounds like a retreat from "mathematics," it is actually the heart of the matter. Cryptography can prove that, *given certain keys and correct verification*, eavesdroppers cannot read or modify messages without detection. It cannot prove that the keys were issued honestly, that the software implementing the protocol has no bugs, or that a government will not compel access. Mathematics gives conditional guarantees. The conditions are where society enters.

This is why privacy on the modern internet has moved toward *zero trust*: a posture in which the network, the intermediary, and even parts of your own infrastructure are treated as potentially hostile by default. Zero trust is not paranoia; it is an acknowledgment of scale. A small network can be guarded by relationships and reputation. A global network cannot. Encryption becomes the last line of defense precisely because earlier lines are so porous. If packets can be redirected, then content must remain safe even on the wrong route. If DNS can be lied about, then the lie must still fail at the cryptographic check. If a Wi-Fi hotspot is malicious, then the session should still be confidential.

For an everyday example, consider the banal act of checking email in a coffee shop. You are surrounded by strangers. The Wi-Fi password may be printed on a chalkboard. The router may be poorly configured. Someone nearby might be running a "twin" network with the same name, hoping your phone will connect automatically. None of this requires sophistication. Yet in a world where your email login used to travel in plaintext, it would have been enough. Encryption changes the baseline. Even if the hotspot is hostile, it cannot simply read your password out of the air. Even if it tries to insert its own fake login page, a correctly implemented browser should refuse the impostor's certificate.

This does not mean you are safe. It means that the attacker's job becomes harder, and harder in a way that is quantified rather than hoped for. That quantification shapes the economics of attack. It is one thing to exploit a common misconfiguration. It is another to break a modern cryptographic primitive. Encryption shifts adversaries toward the weakest link: stolen passwords, compromised devices, social engineering, fraudulent certificates, or coercion.

Coercion is where the political tension around encryption becomes explicit. States have legitimate interests in investigating crime and terrorism; they also have incentives to monitor opponents and control information. Demands for "exceptional access" or "backdoors" are attempts to alter the mathematics: to ensure that someone, somewhere, retains the ability to decrypt. The problem is that cryptographic systems are not like physical locks, where a spare key can be kept in a safe. A backdoor is a structural feature. Once it exists, the question is not only whether it will be used responsibly, but whether it can be secured against theft, replication, or abuse.

The argument often arrives dressed in moderation: surely there is a middle ground between total privacy and total surveillance. The mathematics is less accommodating. A system designed to be decryptable by an authorized third party is also a system designed to be *worth attacking* at that third party. It concentrates risk. It creates a new high-value target: the key escrow, the interception interface, the compliance mechanism. A backdoor intended for lawful use becomes a backdoor that must be defended against unlawful use, and the defender's task is permanent.

Even without backdoors, encryption has limits that are easy to miss because "scrambling" feels total. Encryption protects content, but it often does not protect *metadata*. If you cannot read the letter, you can still observe that a letter was sent, when it was sent, how long it was, and between which addresses. On the internet, these signals can be extraordinarily revealing. A burst of traffic at a certain time can correlate with an event. A pattern of connections can reveal relationships. A regular check-in can reveal routine.

This is where anonymity systems like Tor enter the story. Tor attempts to hide who is talking to whom by routing traffic through a sequence of relays, each of which knows only its immediate neighbors. The idea is to break the easy linkage between source and destination. Yet Tor does not repeal the laws of statistics. An adversary who can observe traffic at multiple points, or who can perform sophisticated correlation, may still infer connections. Privacy becomes probabilistic. The mathematics shifts from "can the attacker decrypt?" to "can the attacker infer?" and inference is a different kind of threat: less absolute, more contextual, and sometimes just as damaging.

If encryption is about protecting messages in transit, there is another frontier: protecting data while it is being *processed*. Much of modern life involves sending data to someone else's computer: a cloud service, a medical platform, a machine-learning model. Even if data is encrypted on the wire, it is often decrypted at the server so computation can occur. That moment of decryption is a moment of vulnerability. It is also a moment of governance: who has access inside the organization, what logs are kept, what subpoenas apply, what insider threats exist.

This motivates one of the most astonishing ideas in modern cryptography: fully homomorphic encryption (FHE). In 2009, Craig Gentry proposed a scheme that, in principle, allows computation on encrypted data without decrypting it. The server performs operations on ciphertexts and produces an encrypted result that only the client can decrypt, as if the server had worked on the plaintext. It is a different vision of privacy: not merely "no one can read my data in transit," but "no one can read my data even while computing on it." In practice, FHE has historically been expensive, and although it has improved dramatically, it remains a specialized tool rather than a universal default. Still, it changes what is imaginable. It turns "trust the server" into "trust the math," at least for certain computations.

Not all privacy problems are about secrecy from outsiders. Some are about limiting what can be learned from *legitimate* access. A hospital needs patient data to provide care. A statistical agency needs census data to allocate resources. A company needs user behavior to improve services. The question becomes: can useful information be extracted without exposing individuals? Here, too, mathematics offers a vocabulary and a set of tools.

One influential approach is differential privacy, introduced by Cynthia Dwork and colleagues in the mid-2000s and developed into a broad framework (as synthesized in the Dwork–Roth monograph). Differential privacy formalizes a simple intuition: the output of an analysis should not depend too much on whether any one person's data is included. If your participation in a database does not significantly change what the system reveals, you are safer. This is not the same as secrecy. It is a bound on influence.

Differential privacy achieves this by adding carefully calibrated noise to results. The calibration depends on the *sensitivity* of the query: how much the answer could change when a single individual's record changes. A count of how many people live in a region has low sensitivity; a query that returns a maximum income has high sensitivity. By adding noise proportional to sensitivity, and by tracking a "privacy budget" parameter often denoted ε (and sometimes δ for approximate variants), the system can offer a guarantee: an attacker, even with auxiliary information, cannot confidently infer whether a particular individual's data was included, beyond a bounded factor.

This guarantee has a sharp beauty, and also a sharp edge. Differential privacy is only as meaningful as the parameters chosen. A very large ε yields little protection. Releasing many statistics consumes privacy budget, because privacy losses compose; a system that answers unlimited questions, even with noise, can be mined until patterns emerge. Differential privacy forces institutions to confront privacy as a finite resource, not a public-relations slogan.

Before differential privacy became prominent, a popular idea in data anonymization was k-anonymity: ensure that each individual in a released dataset is indistinguishable from at least $k - 1$ others with respect to certain identifying attributes. The intuition is straightforward. If you strip names and group people into equivalence classes, you reduce re-identification risk. The problem is that real-world data is rich, and "quasi-identifiers" can be combined in unexpected ways. A dataset may be k-anonymous and still leak sensitive information if everyone in an equivalence class shares a particular attribute, or if an attacker has external data to narrow possibilities. k-anonymity is not useless, but it is not a comprehensive shield. Differential privacy, by contrast, bakes the adversary into the definition.

These techniques—TLS for transport, PKI for identity, Tor for anonymity, differential privacy for statistical release, homomorphic encryption for secure computation—can feel like separate gadgets. They are better understood as responses to a single condition: the loss of reliable boundaries. The internet dissolves boundaries by design. It connects, interposes, caches, reroutes, replicates. That is why it is powerful. It is also why privacy cannot be left to discretion alone.

There is, however, a quieter limit that no amount of cryptography can erase: trust in endpoints. If your device is compromised, encryption can faithfully protect your data from everyone except the attacker who is now sitting inside your device. If a server is compromised, encryption can protect data in transit, but not after decryption unless additional techniques are used. If you voluntarily give your data to a service, cryptography can constrain what the service can do only if the system is designed to make certain misuse impossible, not merely prohibited. Mathematics can create constraints, but only if the system's architecture honors them.

A useful way to think about the mathematics of privacy is as a set of contracts written in a language that computers cannot misunderstand. "Only someone with this key can read this message." "Any single person's data changes the output only a little." "This computation can be performed without revealing inputs." These are extraordinary achievements, but they are also brittle: they require correct implementation, careful key management, sane defaults, and honest governance.

When encryption works well, it fades into the background. You send a message and it arrives. You pay and the payment clears. You download an update and it installs. The greatest compliment to cryptography is that most people do not have to think about it. Yet the reason it can be invisible is that it is uncompromising. It assumes the worst about the world and then builds a small, reliable sanctuary anyway: a channel whose privacy does not depend on the goodwill of the network carrying it.

Chapter 8

The Keepers of the Flame

While the internet appears to be a static utility, it is actually a chaotic ballet kept in motion by exhausted humans and fragile agreements. We peel back the interface to reveal the constant battle against entropy, the 3 A.M. alerts that prevent global silence, and the invisible parliament that steers the future of our digital nervous system.

8.1 The 3 A.M. Pager

At three in the morning, the world's most powerful computers often look less like science fiction and more like plumbing.

A modern internet service is expected to behave like a utility: you tap an icon, and something arrives. A map appears. A payment clears. A message slips across continents. The expectation is so steady that even a few seconds of delay can feel like a personal insult. Yet the machinery underneath is not a single machine. It is a restless colony of machines: computers in data centers, cables under streets and oceans, routers making per-packet decisions, caches in anonymous closets, radios on towers, batteries in basements, and software processes that appear and vanish with the tide of demand. The whole assemblage works only because it is watched.

The watching is not romantic. It is screens and numbers, alarms and checklists. It is the professional habit of assuming that anything that can fail eventually will, and that failure rarely announces itself politely. The internet is not self-healing in the way skin is self-healing. A cut in skin triggers inflammation, clotting, tissue growth. A cut in the internet triggers nothing but the consequences of being cut: requests time out, retries pile up, queues overflow, caches serve stale data, customers refresh, other systems slow down in sympathy. Some components do recover automatically, and engineers have built many clever mechanisms that re-route around trouble. But those mechanisms are not nature; they are policy, code, and often compromise. The difference between an inconvenience and a catastrophe is frequently a person waking up, understanding what is happening, and doing something purposeful before the situation hardens into a larger failure.

The ritual of being that person has a name: *on-call.* Somewhere, in a quiet apartment lit by a charging phone, an engineer has agreed that for a week, or for a night, they will be reachable and ready. The contract is not simply availability. It is readiness to respond with appropriate urgency: to decide whether an alarm is real, to judge severity, to pull in help, to apply a fix or a mitigation, to communicate outward, and to leave behind a trail that makes the next incident less likely. That trail matters because a system that demands heroics every night is not reliable; it is merely *surviving.*

The phrase "pager" survives from an earlier era, when a plastic device on a belt would buzz and display a number to call back. Many teams now use phones and chat apps, but the cultural meaning is unchanged. A "page" is the insistence that sleep is negotiable and impact is not. You can hear it in the language of operations: *acknowledge*, *escalate*, *mitigate*, *resolve*. Even the verbs are sharp-edged.

On-call work exists because the internet is built from layers that conceal their fragility until they cannot. Consider what a simple action really asks for: opening a news site. A device asks a resolver for a name; the Domain Name System (DNS) answers with an address. A connection is negotiated, packets are routed across networks whose business relationships you do not see, and the service responds from whichever data center is "closest" according to a constantly changing map. The response may depend on cached content, on a database lookup, on an ad auction, on an authentication token, on a recommendation engine. Any link can be slow, wrong, overloaded, or misconfigured. The user sees only a spinning icon. The people on-call see a sudden bend in a graph.

Those graphs are the engineer's weather. They are called *observability* tools: dashboards, logs, traces, synthetic probes, and alerting systems. "Observability" is a fashionable word for an old need: to infer what is happening inside a complex machine by looking at its exhaust. A single chart rarely tells the full story, because modern failures are rarely single. A router drops a class of packets, which causes some requests to retry, which increases load, which makes a database fall behind, which increases latency, which triggers a circuit breaker, which changes the traffic pattern, which makes caches less effective, which increases load again. From a distance it can look like a tantrum. From close up, it looks like a set of feedback loops.

Operations centers formalize that closeness. A Network Operations Center (NOC) is part control room, part customer service, part emergency dispatch. In a large organization the NOC may be the first to notice that something odd is happening across a fleet, especially when the symptoms appear in multiple services. In smaller organizations the NOC is virtual: a rotating cast of people watching dashboards and responding in a shared chat room. Either way, the point is to compress time. The earlier you notice, the more options you have.

A page at 3 A.M. is often triggered by an automated threshold: error rate above $x\%$, latency above y milliseconds, a queue length climbing too fast, a data center losing contact, disk space running low in a way that indicates a runaway log. Those thresholds are crude by necessity. A system that paged for every minor anomaly would destroy the people it depends on. A system that pages only when catastrophe is undeniable will page too late. Finding the balance is a form of engineering, but also a form of anthropology: it requires understanding what humans can sustain.

The critical moment, when a machine's observation becomes a human obligation, is the moment the internet stops being a story about technology and becomes a story about labor.

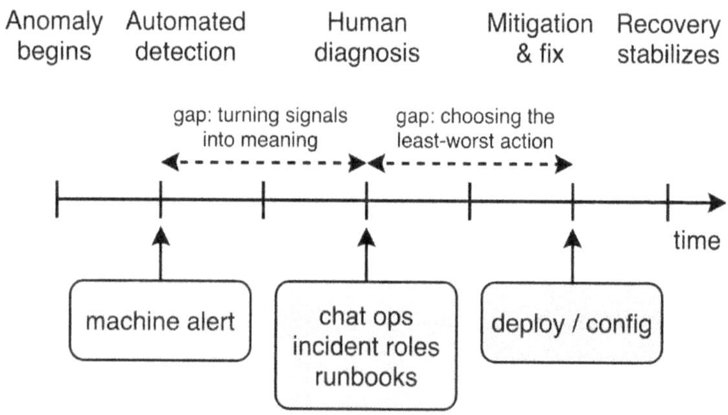

The timeline hints at a truth that people outside operations often miss: "time to detect" is only the beginning. What matters is "time to understand" and "time to act," and those are stubbornly human.

A concrete incident tends to begin with something small, almost embarrassing. A certificate expires. A script that rotates logs silently fails after an operating system update. A network change is applied in the wrong place. The modern internet has made changing systems easy, which is another way of saying it has made introducing new mistakes easy. If you can deploy software fifty times a day, you can deploy a mistake fifty times a day.

One of the most widely discussed outages of recent years, because it was both dramatic and strangely clean, happened to Facebook on October 4, 2021. For hours, Facebook's services were largely unreachable, and the disruption rippled outward: messaging failed, internal tools were harder to access, and users saw the unsettling feeling of a major platform simply not being there. Public reporting indicated that a network configuration change affected the routing information Facebook announced to the rest of the internet, making its systems difficult to reach. The episode was a reminder that the internet's connectivity is, in practice, a constant conversation among routers, and that a single organization can accidentally step out of that conversation. It also illustrated an operational cruelty: when your authentication systems and internal documentation are themselves part of the same network fabric, an outage can make the tools for fixing the outage harder to reach. The engineers fixing the problem were not fighting a single broken server; they were fighting the fact that their own map had gone dark.

At 3 A.M., outages are usually less cinematic but more psychologically demanding. The alarm arrives without context. The first task is triage: *Is this real? Is it urgent? How many people are affected?* Severity is not just a moral judgment; it is a coordination tool. High-severity incidents require forming a team quickly, because attention is a scarce resource and confusion is contagious. Many organizations assign explicit roles in the heat of an incident: someone to manage the response and keep the group focused on restoring service, someone to record events and decisions, someone to handle communications so that the people changing systems are not also crafting careful updates. That separation matters because stress narrows thinking. Under pressure, even experienced engineers can fixate on their favorite theory, or chase symptoms, or forget to say what they are doing. Clear roles are an antidote to the cognitive tunnel.

A typical incident begins in a chat room that becomes, for a few hours, a temporary operations center. The chat fills with clipped statements:

Errors spiking in us-east.

Seeing elevated latency on checkout.

Rolling back last deploy.

Anyone from database on?

Customer support says payments failing in Europe.

The language is terse because the medium rewards terseness, but also because people are trying to keep their own minds from scattering. The scribe timestamps key actions. The incident manager asks for periodic summaries: what is known, what is being tried, what is the next decision point. The most valuable sentence in such a chat is often the simplest: *I don't know yet.*

The tools that feed this response are designed to make an invisible world navigable. Logs are the system's diary, but at large scale they are less a diary than a landfill, and searching them is an art. Metrics provide a kind of vital sign: requests per second, error rate, latency percentiles, saturation of CPU or memory, queue depths. Traces follow a single request through many services, like tracking a single raindrop through a city's drainage system. Each tool answers a different question. Metrics tell you *that* something is wrong; logs and traces help you decide *why*. None of them is sufficient alone.

The act of diagnosis is often a matter of translating between layers. An alert says "HTTP 500 errors increased." That is not a cause; it is a symptom. The cause might be that a database is slow, which itself might be because a disk is failing, which might be because a controller firmware bug is triggered by a specific workload pattern, which might have been introduced by yesterday's "small" change. The chain of causality crosses software and hardware, networks and power, human decisions and automated safeguards.

When the diagnosis is uncertain, the operational instinct is to buy time. Mitigation is not the same as fixing the root cause. It might mean reducing load by turning off a nonessential feature, shedding traffic, serving cached results, temporarily raising a rate limit, or rerouting requests away from a troubled region. A mitigation can be ugly, but it keeps the patient alive while the doctors argue. Many outages are shortened not by brilliance, but by choosing a reversible action that reduces harm.

Reversibility is a central value in on-call culture because the internet is too complex to fully understand in real time. Rolling back a deploy is popular not because it is always correct, but because it is often the fastest way to return to a previously known state. Yet rollback is not a magic button. Data migrations, cache warmups, and stateful changes can make "going back" expensive or impossible. Even a simple rollback can fail if the rollback path has not been tested recently, or if dependencies have moved on. The cruelest discoveries at 3 A.M. are the ones that reveal a comforting assumption to be false: that a standby system is actually ready, that failover actually works, that yesterday's change can be undone cleanly.

The most difficult part of the work is not typing commands. It is managing attention across many signals while sleep-deprived. This is where the notion of *toil* becomes relevant. Toil is operational work that is manual, repetitive, reactive, and has no enduring value: the same alert every night, the same dashboard check, the same *just restart it* ritual. Toil is dangerous not only because it wastes time, but because it erodes judgment. When an engineer has been paged a hundred times for trivialities, the hundred-and-first page may be ignored, and the one time it mattered will arrive wrapped in the fatigue created by the other ninety-nine.

Alert fatigue is not merely an individual weakness; it is a system property. Poorly designed alerts create noise, and noise creates indifference. A good alert is actionable: it points to a condition that requires human intervention, and it arrives with enough context to make the first decision quickly. Many teams revise alerts the way cities revise traffic lights: gradually, after near-misses, trying to shape behavior without causing new jams. The hidden sophistication of a mature operations practice is not in its hero stories, but in the quiet reduction of needless interruptions.

The geography of the internet also shapes the culture of response. Data centers are physical places with physical constraints. Even in a world of remote work and cloud abstractions, there are moments when a problem is stubbornly tactile. A fiber cut in a construction zone is repaired by people with shovels and splicing equipment. A failing switch might require a technician to swap hardware. A cooling issue can turn into an emergency if temperatures rise and servers throttle or shut down. Power is a constant background concern; the most elegant software cannot compute in the dark. Many operators learn to keep an eye on local weather not out of curiosity but because storms and heat waves have predictable effects on grids and on cooling systems. The internet's nervous system is powered by other nervous systems.

Because the internet is a patchwork of organizations, a single incident can require diplomacy. An outage might be caused by a transit provider, by a DNS operator, by a certificate authority, by a cloud region, by a content delivery network. Escalation sometimes means calling another company's on-call engineer. In that moment, the internet looks less like a unified technology and more like a web of mutual aid agreements, service-level contracts, and professional courtesy. If you have ever wondered why a status page update can sound cautious or vague, part of the reason is that the truth may depend on what another organization is still investigating.

A less obvious human task is communication with people who are not engineers. When a service fails, customers want to know whether their data is safe, whether they should retry, whether they should do something differently. Colleagues want to know whether to pause deployments, whether to reroute traffic, whether to expect degraded performance. Executives want timelines and impact estimates, often before those can be known reliably. Good incident communication is not spin; it is a way of reducing secondary damage. In many industries, confusion is costlier than the initial outage because it leads to duplicate work, contradictory actions, and rumor.

Once the crisis passes, the temptation is to forget. Sleep is reclaimed. Dashboards return to their calm baseline. The best operations cultures resist forgetting, not out of masochism, but because memory is a form of prevention. A post-incident review turns a stressful night into a permanent change: a missing alert is added, a brittle dependency is removed, a runbook is improved, a risky procedure is automated, a single point of failure is given redundancy, a confusing dashboard is redesigned so that the next person does not have to guess.

These reviews are most useful when they are honest about human factors. People make mistakes under pressure; the goal is to design systems that are resilient to ordinary human fallibility. If a single mistyped command can take down a critical service, that is not merely a personal failure waiting to happen; it is a design that treats perfection as a prerequisite. At 3 A.M., perfection is not available.

The internet's apparent smoothness is therefore a kind of social achievement. It is maintained by a dispersed guild of operators who accept that a portion of their working lives will be interrupted, that they will think clearly while tired, that they will make decisions with incomplete information, and that they will build tools to make those decisions less punishing over time. The technology matters immensely, but it is shaped around a human reality: complex systems do not stay healthy by accident. They stay healthy because someone is listening for the first faint change in tone, and is prepared to act before the rest of the world hears it.

8.2 Fighting the Red Queen

The most expensive work on the internet rarely looks like innovation. It looks like replacement.

A new feature is easy to photograph. A redesign can be announced. A product launch has a date, a logo, a stage. Maintenance has none of these. It is the work that must happen for a service to remain boring, and "boring" is the highest compliment a network can earn. The strange thing about the internet is that its primary activity is not invention but upkeep: patching old code, rotating certificates, renewing domain registrations, replacing failing disks, migrating away from systems that are still working but no longer safe, and revisiting decisions that were sensible when traffic was a tenth of what it is now. If the internet feels like it is racing forward, that is partly because it must run hard simply to avoid drifting backward.

Biologists call this kind of predicament the Red Queen effect, after the character in Lewis Carroll who must keep running to stay in the same place. In evolutionary terms, the environment is not fixed. Predators adapt, prey adapts, parasites adapt, hosts adapt. Success is not arriving at stability; it is maintaining relative fitness in a world where everyone else is also changing. The internet has its own Red Queen, and she is not a single adversary. She is time, scale, entropy, and other people's updates.

Time is corrosive. Hardware ages in predictable ways: fans wear out, power supplies fail, solder joints degrade, spinning disks die, and even solid-state drives have finite write endurance. Networks built from optics depend on clean connectors and stable amplification. Power systems depend on batteries that lose capacity and generators that must be exercised like muscles. These are not exotic failures; they are the dull physics of materials. A data center is a machine made of other machines, and every component has a curve of declining reliability.

Scale is corrosive too. A pattern of failure that happens once a year on a hundred servers can happen daily on ten thousand. A small percentage of user requests failing can still mean thousands of people at any moment. At large scale, "rare" becomes a constant background hum.

Entropy in software is less visible but no less real. Code depends on other code: libraries, compilers, operating systems, databases, container runtimes, drivers, and the endless substructure that most applications no longer build themselves. Those dependencies update. They deprecate old behavior. They fix security bugs by tightening defaults. They remove features that seemed essential five years ago. A system that never changes is not "stable"; it is increasingly out of step with the world that surrounds it.

Then there are the changes that arrive with agency: attackers probing, competitors optimizing, customers behaving in new ways, regulators demanding new controls, browsers enforcing new security requirements, and large providers shifting their own policies. Even well-meaning changes in one corner of the ecosystem can break assumptions elsewhere. The internet is a network of networks and a stack of stacks; no one is in charge of the whole pile.

Maintenance, under these conditions, is not a side task. It is the main task. It is the ongoing effort to keep a complex system within the narrow band where it remains useful.

The labor of on-call, with its alarms and rapid triage, is the visible tip of that work. The deeper part is what follows once the immediate emergency passes: the intentional reduction of the number of future emergencies. This is where operational culture becomes a kind of ethics. Teams can respond to outages as isolated dramas, each one solved and forgotten, or they can treat each incident as evidence that the system is mis-shapen, demanding corrections that make the next incident less likely or less severe.

Engineers sometimes call the repetitive, interrupt-driven part of operations *toil*: tasks that are manual, predictable, and reactive, that provide no enduring value and scale linearly with the system. Toil is not necessarily low-status work; it is often the work that keeps services alive. But it is dangerous when it becomes the majority of what a team does, because a team drowning in toil has no time to build the improvements that would reduce the drowning. A service can become a kind of treadmill: each week's growth produces more alerts, which consumes more attention, which delays the automation that would have prevented the alerts, which produces even more alerts.

A striking feature of mature reliability practice is how explicitly it tries to break that loop. Organizations that take uptime seriously often protect time for maintenance as fiercely as they protect time for new features, because reliability is not something you "add" later. It is something you *maintain* continuously, like cleanliness in a kitchen. Skip it for a month and you do not merely accrue dirt; you change what kinds of problems become likely. Fires start more easily.

The internet's Red Queen is perhaps most obvious in cybersecurity, where the adversary actively adapts. A security team that stops patching does not become static; it becomes vulnerable in new ways each week as attackers incorporate newly published weaknesses into their toolkits. Defensive measures that worked last year can become predictable and therefore less useful. The paradox is that even when defenders are doing well, the reward is invisibility. A prevented breach is not a headline. A patched library that quietly closes a critical hole is rarely celebrated outside the team that applied it.

Yet maintenance is not only about defending against attackers. It is also about defending against your own past. Early decisions are made under pressure, with incomplete information, and with the seductive optimism of a system that has not yet met its full scale. The internet is full of clever shortcuts that worked until they didn't: a database table that seemed comfortably sized, a caching strategy that assumed traffic patterns would remain similar, a routing policy that relied on informal trust, an alert that assumed failures would be loud rather than subtle. As systems grow, yesterday's shortcut becomes today's constraint.

The name for this accumulation is *technical debt*. The metaphor is financial: you take on debt to move faster now, but you pay interest later in the form of complexity, fragility, and maintenance cost. The metaphor is imperfect, because technical debt is not always a sin. Sometimes it is the right trade. A service that never takes shortcuts may never ship. But interest compounds. A system with too much debt becomes brittle. Every change becomes risky, because there are too many unknown dependencies and too many hidden assumptions. Teams begin to fear deploying. They schedule changes only during narrow windows. They rely on a few veterans who remember why things are the way they are. Those veterans become single points of failure in human form.

The Red Queen effect turns technical debt into a trap. The world changes whether you refactor or not. If you delay maintenance, you are not choosing to "stay the same." You are choosing to let the gap between your system and its environment widen until it becomes painful.

A concrete example helps. In August 2014, a security vulnerability in OpenSSL, a widely used cryptographic library, was disclosed: the bug became known as *Heartbleed.* The details are less important than the lesson. OpenSSL was embedded everywhere: in web servers, VPNs, appliances, and systems whose owners might not even have realized they were depending on it. When Heartbleed became public, the internet did not "heal." Administrators had to patch. Certificates had to be replaced. Passwords had to be changed. Some organizations moved quickly; others took weeks or months. Some systems could not be patched easily because they were embedded devices with awkward update mechanisms, or because the software had been customized and the maintainers were gone.

Heartbleed was a moment when maintenance became visible to the public: not as a routine, but as a scramble. It also exposed a structural weakness of the modern internet: foundational components can be maintained by surprisingly small groups of people, and yet be depended upon by the entire world. A library can be both ubiquitous and underfunded. The internet's smooth surface sometimes rests on a few volunteers and a small donation page.

Maintenance is also a story about *coordination*. A single engineer can patch a server. But many threats to stability are emergent properties of the network: they arise from how systems interact. Routing is an especially sharp example, because routing determines what is reachable at all.

When a packet moves across the internet, it does not follow a globally computed optimal path. It follows a series of local decisions made by autonomous networks, each with its own business relationships and policies. The protocol that distributes reachability information between these networks is called the Border Gateway Protocol, or *BGP*. BGP is a triumph of pragmatism: it allows thousands of independently operated networks to exchange enough information to make global connectivity work. It is also, by design, trusting. A network can announce, "I can deliver traffic for these addresses," and its neighbors may believe it, propagate that belief, and send traffic accordingly.

Most of the time, that trust is not abused. But the Red Queen effect shows up in the gap between "most of the time" and "always." A simple configuration mistake can become a *route leak*: a network unintentionally announces routes beyond the scope it should, persuading others to send traffic in a way that violates intended policy. Because the internet's interconnections are shaped by commercial relationships (who pays whom for transit, who peers for free), route leaks are not just technical errors; they can scramble the economic assumptions that keep traffic flowing smoothly. A leak can attract traffic to a network that cannot handle it, producing congestion and outages far from the source of the mistake.

These events are sometimes reported in the press as if they were mysterious acts, but they are often mundane misconfigurations amplified by the internet's size. The maintenance burden here is not merely "keep routers running." It is continuously revisiting filters, policies, and monitoring so that an accidental announcement does not spread unchecked. It is building systems that can detect abnormal route changes quickly and either alert humans or automatically dampen the impact. It is agreeing, across organizations, on practices that make the overall system more robust. Unilateral perfection is not enough; your neighbor's mistake can become your outage.

This is where the Red Queen becomes not just a metaphor but a management problem. An operator can do everything "right" internally and still suffer because the surrounding ecosystem is changing. So maintenance expands outward: participation in operational communities, adoption of best practices, shared tooling, and standards efforts that recognize that routing security and stability depend on mutual checks. The internet's resilience is not solely a product of clever protocols. It is also a product of a culture in which operators compare notes, warn one another, and slowly ratchet up expectations.

The physical world imposes its own form of running. Undersea cables are marvels of engineering and a reminder that "wireless" is mostly marketing. Those cables are repaired by specialized ships that retrieve a cable from the sea floor, cut it, splice in a new segment, and lower it back down. Even when the repair is successful, the outage has already happened. The maintenance work in this domain often looks like redundancy planning: multiple routes, diverse landing stations, agreements with alternative carriers, and constant testing of failover paths. Redundancy is expensive, and the pressure to reduce cost is relentless, which means maintenance is always negotiating with finance. Reliability is not simply a technical property; it is an economic choice made repeatedly.

A similar negotiation plays out in software. Engineers want to keep dependencies current, because older versions accumulate security vulnerabilities and become harder to support. Product teams want features. Customers want new capabilities and lower prices. Maintenance competes for attention, and attention is the scarce resource from which outages are made.

At its best, maintenance is not defensive crouching but active cultivation. A well-maintained system becomes easier to change, not harder. Documentation is kept close to the code. Runbooks are updated after incidents. Backups are tested rather than assumed. Capacity is planned with headroom. Automated tests catch regressions before they reach production. Alerts are tuned so that a page usually means "wake up and act," not "wake up and suffer." Teams practice incident response so that the first time they coordinate under pressure is not in front of a live audience of millions.

At its worst, maintenance becomes superstition: rituals that are performed because they have always been performed, not because anyone is sure they still matter. This is another kind of debt, a debt of understanding. People reboot servers because it "usually helps," without knowing whether they are clearing a memory leak or merely masking a deeper bug. They avoid upgrading a library because the last upgrade was painful, even though the pain is precisely what compounds when upgrades are delayed. They treat parts of the system as untouchable, and those parts become the parts most likely to fail spectacularly.

The Red Queen does not punish ignorance out of malice. She punishes it mechanically. Complex systems drift. Assumptions expire. The things you do not understand become the things you cannot safely change, and the things you cannot safely change become the things that trap you when the world forces change anyway.

There is a quiet heroism in the people who choose to do this work well, because the reward is not applause but the absence of disaster. Their victories are negative space: no outage today, no breach this week, no cascading failure during a traffic spike. Even internally, good maintenance can be hard to defend. It is easier to ask for funding for a new product than for replacing a working router whose support contract ends next year. It is easier to justify a feature that users will notice than a refactor that only engineers will feel.

And yet the internet is an environment where the cost of neglect is not merely technical. It is social. When networks fail, appointments are missed, wages are delayed, emergency information is harder to reach, and trust in institutions erodes a little more. The internet has become a civic substrate: informal, unevenly governed, but deeply entwined with ordinary life. Maintenance, then, is not merely the avoidance of inconvenience. It is the continuous work of keeping a shared nervous system from misfiring.

Running to stay in place can sound bleak. It need not be. There is a craft to it, and a certain satisfaction in turning a fragile contraption into something that can absorb shocks without drama. A patched system is not merely safer; it is more honest. It acknowledges that change is constant and that reliability is an achievement, not a default.

On a quiet weekday afternoon, an engineer merges a change that updates a dependency no user has heard of. A technician replaces a power supply in a rack that has never been mentioned in a meeting. An operator tightens a routing filter that prevents an accidental announcement from escaping into the wider world. None of this will trend. None of it will be celebrated. And because it happened, the internet remains, for another day, ordinary.

8.3 The Bazaar and the Boardroom

People often talk about “the internet” as if it were a single thing with a single owner, like a railroad or a power grid. In practice it behaves more like a city built by committees that disagree on almost everything except the need for roads. There are building codes, but they are voluntary. There are police, but their jurisdiction stops at the property line. There are neighborhood associations, but many residents never attend meetings. And yet the traffic mostly flows.

That apparent contradiction is the internet's governance: a paradoxical mix of volunteer cooperation and hard-nosed economics. On one side is the bazaar, where engineers argue in public, trade drafts, write open-source code, and advance proposals through rough consensus. On the other side is the boardroom, where networks sign contracts, negotiate peering, buy transit, and sometimes decide—quietly—that one link is not worth the cost of another. The miracle is not that the internet is free of politics. The miracle is that it functions while saturated with them.

The phrase "rough consensus" comes from the culture of the Internet Engineering Task Force (IETF), the open standards body that has shaped much of the internet's basic machinery. The IETF is not a parliament and not a regulator. It does not "pass laws" in the ordinary sense. It produces documents-Requests for Comments, or RFCs-that describe protocols and best practices. Some RFCs are informational, some experimental, some become internet standards. The IETF's authority is a kind of earned legitimacy: it matters because people implement what it writes.

Rough consensus is often misunderstood as hand-waving or a popularity contest. It is more like a refusal to treat procedure as a substitute for persuasion. The goal is not unanimity; it is to prevent small factions from blocking progress while still taking objections seriously. In the IETF's self-image, arguments are supposed to be grounded in engineering reality. The phrase "running code" captures that attitude: a proposal that can be implemented and tested carries a weight that a purely rhetorical position does not. This is the bazaar at its best—messy, argumentative, but oriented toward building things that work across many vendors and many networks.

Eric S. Raymond's essay "The Cathedral and the Bazaar" (1998) gave this style of development a memorable contrast. The cathedral is planned, controlled, released in carefully managed versions. The bazaar is open, iterative, full of simultaneous experiments. The internet contains both. Many of its most influential building blocks-TCP/IP, DNS, BGP, HTTP-were standardized through a bazaar-like process of drafts, debate, and implementation. Yet the internet's largest services and networks are run like cathedrals: tightly managed, optimized for scale, and guarded as competitive advantage. Even within a single company, the tension persists: engineers may advocate for open standards and interoperable designs while executives push for lock-in, differentiation, and margin.

The internet's governance becomes easier to understand if it is separated into two kinds of problems. One kind is *coordination*: agreeing on technical conventions so that systems can talk to one another. The other kind is *allocation*: deciding who pays for what, who connects to whom, and under what terms. The bazaar is good at coordination. The boardroom is good at allocation. Neither is sufficient alone.

Consider naming, the quiet miracle that lets you type a human-friendly address rather than memorizing numbers. The Domain Name System is not merely a protocol. It is also an administrative hierarchy. Someone must decide what counts as a top-level domain (`.com`, `.org`, `.uk`), who operates it, and how changes are made without breaking the world's ability to find things. That "someone" is not a single sovereign. It is a set of institutions that coordinate the internet's unique identifiers: numbers, names, and protocol parameters.

One of the most concrete choke points is the DNS root zone: the small, intensely important database that lists top-level domains and points to the servers that answer for them. The technical work of maintaining that root zone is handled by the Internet Assigned Numbers Authority (IANA), and the broader coordination is associated with the Internet Corporation for Assigned Names and Numbers (ICANN). The structure is strange by the standards of nation-states: ICANN describes its policy development as bottom-up and multistakeholder, involving governments, businesses, civil society, and technical community participants, while its Board retains ultimate approval authority. This is neither a pure democracy nor a pure corporation; it is an engineered compromise, designed to be legitimate enough to be accepted and structured enough to be operational.

That compromise was stress-tested by the IANA stewardship transition. For decades, the U.S. government—through the National Telecommunications and Information Administration (NTIA)—had a stewardship role over the IANA functions contract. Critics saw this as an anachronism: a global network with a uniquely American oversight hook. Supporters saw it as a stabilizing backstop. In 2014, NTIA announced its intent to transition stewardship to the global multistakeholder community, and on October 1, 2016, ICANN stated that the contract had expired and the transition had taken effect, accompanied by new accountability measures. The technical effect for most users was invisible. The political symbolism was enormous. It was one of the rare moments when the internet's governance machinery briefly surfaced into mainstream attention, not because a website went down, but because the question "Who ultimately controls the root?" could not be ignored.

For everyday internet stability, however, the more relentless form of governance is not about names. It is about how traffic moves and who pays. This is where the boardroom dominates.

Every time you watch a video, send a photo, or load a page, packets cross multiple independently operated networks. These networks are called Autonomous Systems (ASes), and each one has its own priorities. Some are retail internet service providers (ISPs) that sell access to homes and businesses. Some are content providers with vast data centers. Some are content delivery networks (CDNs) that place caches near users. Some are backbone carriers that specialize in moving traffic long distances. They connect through a web of agreements that fall, broadly, into two categories: *peering* and *transit*.

Peering is a bargain between equals, or at least between parties who agree to treat one another as equals for a specific purpose. Two networks connect and exchange traffic destined for each other's customers. Often no money changes hands; the benefit is mutual. Peering can be "settlement-free" (no payments) or "paid peering" (payments exist but the relationship is still framed as an exchange rather than a wholesale service).

Transit is different. Transit is the purchase of reachability. If a small network buys transit from a larger one, it is effectively paying to have its traffic carried to the rest of the internet and to have the rest of the internet's traffic carried back. Transit is a service, sold under contract, usually with performance commitments and consequences for noncompliance. The customer pays. The provider delivers.

These distinctions can sound abstract until one asks a blunt question: why does the internet still work if there is no central authority? One answer is that money creates incentives strong enough to hold the system together. Contracts substitute for law across borders. If you do not deliver packets reliably, your customer can leave. If you cannot attract customers, you cannot pay for capacity. If you cannot pay for capacity, you cannot compete. Governance emerges from competitive pressure as much as from technical ideals.

Yet that same economics also creates friction, and friction becomes outages in the real world. When two networks argue about peering terms, users can experience slowdowns or disconnections even though nothing “broke” in the technical sense. Congestion at an interconnection point can be as damaging as a failed router. A network may decline to upgrade a link if the other side refuses to share costs. A content provider may decide that it is cheaper to build its own private backbone than to keep paying transit. A broadband ISP may decide that its leverage lies in scarcity at the edges. These are not engineering disagreements; they are business negotiations carried by fiber.

The economic hierarchy of the internet can be sketched with a simple picture. At the top sit so-called Tier 1 networks, carriers that can reach every other network on the internet without buying transit, because they peer with one another. Below are regional providers, ISPs, enterprises, and content networks that buy transit or peer selectively. The structure is not perfectly stable—markets change, mergers happen, new players build backbones—but the underlying idea persists: someone pays someone else to carry bits.

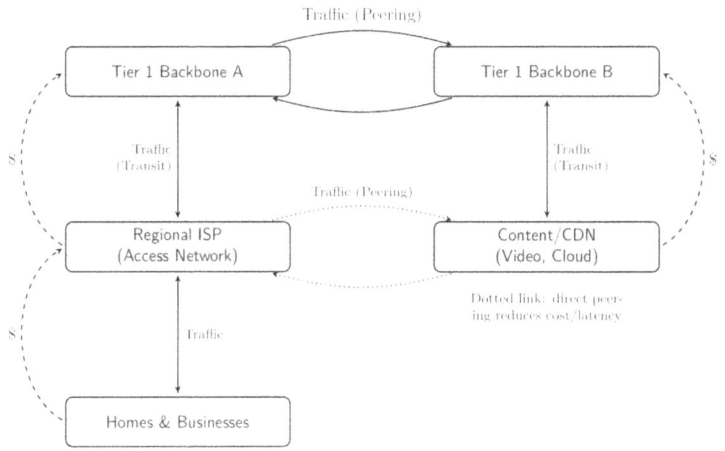

The diagram is simplified, but it captures a central point: there are two flows to keep separate in one's mind. Data flows in both directions, because communication is reciprocal. Money flows in specific directions, because economics is asymmetric.

This economic governance shapes technical choices in subtle ways. A network may prefer a route that is not the shortest, but the cheapest under its contracts. A provider may place caches in a partner's facilities to reduce transit costs and improve performance. A content company may build a private backbone to avoid being squeezed by transit pricing. The topology of the internet is not only a matter of physics; it is also a map of invoices.

The bazaar enters again when those economic realities threaten the health of shared protocols. Routing security provides an example. BGP's trust model worked tolerably well when the network community was small and relatively collegial. As the internet expanded, accidental route leaks and deliberate hijacks became more significant. The technical community has proposed and developed mechanisms to reduce the damage, including practices and standards that rely on shared validation. The difficulty is that many improvements have the character of public goods: they cost money and effort for one operator, while the benefits spill outward. If your neighbor does not participate, your investment may not fully protect you. This is a governance problem disguised as a protocol problem.

The boardroom also intrudes on the bazaar in another way: deployment incentives. The IETF can publish a protocol that improves security or performance, but adoption is voluntary. If the protocol requires replacing equipment or changing business processes, some operators will delay. If the benefits accrue mostly to others, adoption can stall. The internet, in other words, is not governed by the elegance of its standards. It is governed by the incentives that make those standards worth implementing.

And yet the bazaar has a kind of stubborn power. It produces shared vocabularies and shared expectations. It provides a venue where engineers from rival companies can cooperate on the unglamorous machinery that everyone needs. It also creates a record: written protocols that outlast product cycles and executive reshuffles. When the bazaar works well, it is a slow form of diplomacy conducted in packet formats and failure modes.

What makes this blend paradoxical is that it produces both openness and concentration. The protocols are public, and in principle anyone can implement them. But the economics of scale concentrate power. A few large platforms can build global infrastructures that smaller organizations cannot match. A few Tier 1 carriers can shape the backbone. A few DNS and CDN operators can become critical intermediaries. The internet remains interoperable, but not necessarily egalitarian.

One might expect such a system to collapse under mistrust. Instead it muddles through, because mistrust is managed rather than eliminated. Networks authenticate each other through contracts, communities of practice, and an accumulation of defensive mechanisms. Standards aim to reduce ambiguity. Operators build monitoring to detect when the outside world is behaving strangely. When relationships fail, traffic is rerouted, lawsuits are filed, customers churn, and the system reconfigures. Governance happens not only in meetings, but in the daily act of keeping packets moving in a way that is profitable enough to sustain the infrastructure.

It is tempting to look for a single lever—some global regulator, some technical silver bullet—that could make the internet's governance clean. The internet resists cleanliness. It is too large, too international, too commercially embedded, and too technically diverse. Its stability comes from a patchwork of institutions and habits: engineers arguing about formats and timers, lawyers negotiating contracts, operators setting filters, and companies betting billions on cables and data centers that will be obsolete before they are fully depreciated.

The result is a system that behaves, at its best, like a commons maintained by people with very different motives. A volunteer can propose an improvement to a protocol because it seems like the right thing to do. A corporation can adopt it because it reduces cost or risk. A regulator can reference it because it has become a de facto standard. The same packet can travel under the moral logic of cooperation and the financial logic of profit, and arrive without caring which motive was stronger.

Late at night, when an outage spreads across networks and the on-call engineers begin calling their counterparts at other companies, the bazaar and the boardroom meet in a single exhausted sentence: "We think the problem is at the interconnect." If they are lucky, the contract specifies who must fix it. If they are very lucky, the engineers on both sides share enough trust to work quickly even while the lawyers sleep. The internet remains alive in that narrow overlap between common purpose and negotiated self-interest, where the fastest path to fixing a shared problem is still to talk to the other humans who keep the lights on.

www.ingramcontent.com/pod-product-compliance
Lightning Source LLC
LaVergne TN
LVHW022209200926
848481LV00006B/1901
9798896652694